HEALTHFUL HOUSES

HOW TO DESIGN AND BUILD YOUR OWN

Clint Good, AIA-Architect
Debra Lynn Dadd

GUARANTY PRESS
Bethesda, Maryland

Good, Clint
Dadd, Debra Lynn
Healthful Houses:
How to Design and Build Your Own

ISBN:
0-9620892-0-6

Library of Congress Card Number:
88-081856

Cover photograph by:
Neil Steinberg, Leesburg, Virginia

Requests for such permission should be addressed to:
Guaranty Press
7315 Wisconsin Avenue
Suite 615E
Bethesda, MD 20814
1-800-541-9185

ACKNOWLEDGEMENTS

I would like to thank Debra Lynn Dadd for her assistance in producing this book. Debra is fully aware of the importance of man's unity with nature and is sincere about promoting a lifestyle that recognizes this balance.

Special thanks to Ellen Flynn-Heapes and Kristen Coffield of Flynn-Heapes Consulting in Alexandria, Virginia, and Lori Bills-Wiseman Design in Falls Church, Virginia.

And finally, I wish to dedicate this book to my wife Mary Ann, whose love for nature is most supportive of my work.

TABLE OF CONTENTS

From The Authors and Publisher . . .

The information in this book is intended for both the general public and building professionals interested in a building project. The information is not intended to give advice with respect to any medical condition, nor do the authors or publisher guarantee any of the products mentioned in this book to be tolerable for all individuals, regardless of their level of sensitivity to certain substances.

This book recommends products that can contribute to overall health and well-being by their absence of substances known to be toxic, carcinogenic, mutagenic, or teratogenic. We do not intend to imply, however, that any of these products can in themselves heal or cure illness.

Anyone endeavoring to adjust his environment for the specific purpose of alleviating a medical condition should consult a qualified health care professional.

This book was written by an architect and a consumer advocate with the intention that each reader will take responsibility for the choices of products made.

Despite efforts on our part to be accurate, this book may contain errors. Specifically, we cannot be responsible for:

- incorrect or incomplete information regarding products,
- changes in product ingredients,
- substances or processes with toxicity as yet undiscovered,
- continued availability of any item,
- toxicity of any product listed,
- any adverse health effects caused by using any product.

Writing this book was very important to me because I would like *everybody* to have the opportunity to live in a healthful house. We know that some materials used in modern building methods can be harmful to our health — in recent years architects and builders have become increasingly concerned with indoor air pollution and sick building syndrome.

I believe that we each have the right to live in a home that supports our physical and mental health and well-being. To live in such a home is not just a dream, it is already a reality. This home can be as affordable as any other custom home and built in any architectural style you choose. The key is to concentrate on using the correct materials.

As an architect and builder, I have designed and built many houses. I've followed the tradition of my father and grandfather, both of whom were master builders and craftsmen. Since receiving a formal education, I have been licensed in several states and hold the national certification from the National Council of Architectural Registration Boards (NCARB), as well as membership in the American Institute of Architects (AIA).

My interest in healthful houses began when I became very sensitive to indoor air pollutants and other manmade products of our twentieth century life. Today I am healthy and symptom-free after designing and building a home without the myriad of artificial products that most new homes in America today are made of. I built a house for myself the old-fashioned way, the way I used to build them with my father and grandfather. It was a good experience to build the first house for myself because I learned what to do and what not to do.

Since my first experience, I have built other new "old" houses, all combining modern convenience and new natural materials with the quality of old-fashioned construction methods. There are now many innovative alternative building techniques, but I have chosen to design and build houses that integrate our traditional heritage of home and hearth with our present need for cost-effective and healthful housing, and to focus on materials, methods, and products that can be used to construct a house in any architectural style.

Healthful Houses is written primarily for the health-minded individual building a custom home or remodeling an existing structure, but architects and builders will also find this book a helpful source. Because using healthful materials is such an important part of the design process, this book will focus mainly on the materials themselves, presenting product information in the standard industry format.

Welcome to a whole new world of building possibilities!

Clint Good.

Clint Good

"I welcome all inquiries and would enjoy hearing about your experiences in this whole new world of building possibilities!" Contact: Clint Good, Post Office Box 143, Lincoln, VA 22078.

Designing the Healthful House

This book provides an optimistic overview of the many materials, products, and design techniques that can be used to build a healthful house, which can be applied to any architectural style or budget.

A healthful house is one which provides an indoor environment that contributes to the health and well-being of its occupants. It is a pleasant, functional place, where you can live with the peace-of-mind that you are safe from harm and your basic needs are met.

With this definition, we can easily see that most modern residential construction in America does not result in healthful houses. We are daily barraged by formaldehyde from particleboard, combustion by-products such as sulfur dioxide and carbon monoxide from gas appliances, solvents from paints and finishes, radon from the earth we build on, and many more potentially dangerous pollutants.

Indoor air pollution is known to affect our health by damaging the tissue in our respiratory system and by poisoning our blood. It can cause emphysema, bronchitis, asthma, cancer, birth defects, and increased incidence of upper respiratory disease and heart disease. On a more subtle level, some people experience watery eyes, breathing difficulties, headaches, coughing, frequent upper respiratory problems, aggravation of chronic heart and lung disease, shortened life span, and general poor health. These exposures can also cause a breakdown of the immune system that appears as almost any symptom imaginable. In 1987 the Environmental Protection Agency (EPA) targeted indoor air pollution as a top priority. An earlier study done by the EPA showed that our greatest exposure to toxic substances is right in our homes, even if we happen to live next to a chemical factory.

Scientific studies have established a direct correlation between certain health problems and specific interior pollutants. Let's examine these in more detail:

❐ **Formaldehyde** has many sources in the indoor environment — plywood, particleboard, paneling, urea-formaldehyde foam insulation (UFFI), carpeting, drapes, clothing, indoor combustion sources, and many consumer products. Formaldehyde has been shown to produce nasal bronchial cancer in animals. Studies have suggested increases in cancers of the oral and nasal cavities, pharynx, lung and liver in workers exposed to significant levels. Other complaints linked to formaldehyde include eye irritation, respiratory tract irritation, nausea, headache, rash, tiredness and thirst. Some reproductive disorders have been reported in industrial settings as well. There are many documented cases of bronchial asthma due to formaldehyde sensitization, and exposure to formaldehyde has been demonstrated to produce acute allergic dermatitis in individuals.

❐ **Indoor combustion byproducts** — carbon monoxide and oxides of nitrogen — are present in major concentrations in homes where unvented combustion appliances are used. The worst offenders are those appliances that use some form of gas, but wood stoves are also potential sources of CO_2 as well as side-stream cigarette smoke. The symptoms associated with these pollutants include bronchial constriction, burning eyes, headache, breathing difficulties, and nasal discharge.

❐ **Consumer products** of different sorts produce various toxic reactions, sensitivities and allergic responses, particularly pesticides, cleaning products and many synthetic materials used in interior construction and decorating. Use or misuse of some products has resulted in acute and delayed disorders and even death, and their use therefore has been limited or removed. There is much concern about the use of other widely used products because of incomplete information on their possible risk to health.

❒ **Radon** is a naturally occuring radioactive gas that breaks down into compounds that may cause cancer when large quantities are inhaled over a long period of time. (Radon is discussed in greater detail in a later section.)

❒ **Asbestos** is a naturally occuring mineral that has been used over the years in many building products. The extensive use of asbestos-containing materials in buildings has raised concerns about potential exposure. The EPA estimates that asbestos-containing materials can be found in approximately 31,000 schools and 733,000 other public and commercial buildings in this country. Exposure is known to cause lung cancer and mesotheleoma.

❒ **Airborne biological agents** such as bacteria, viruses, house dust mites, molds, fungi, and pollens can have direct effects provoking infection or allergic responses. A number of contagious diseases are capable of airborne transmission in the indoor environment, such as influenza, measles, mumps, and chicken pox, and an even larger number of microorganisms have been confirmed as producing allergic reactions.

A combination of pollutants may also combine synergistically to produce health effects not seen with any single pollutant, making it extremely difficult to associate health problems with any one particular pollutant that might be found in the home environment. This is particularly true of combinations including formaldehyde, cigarette smoke or radon.

Much of our exposure to indoor pollutants comes from the materials used in the structure itself, especially when they are brand new. In its 1981 report *Indoor Pollutants*, the National Research Council recognized the importance of using healthful materials by stating that architects should "treat indoor environmental quality as a design objective . . . there are many opportunities in building design to separate people from the sources of contamination or to remove the sources entirely."

Fortunately, you *can* reduce your exposure to these potential hazards. Whether you are building a home from the ground up or doing remodeling, there are many new products and time-proven methods which can be used that both contribute to your overall well-being and have plenty of aesthetic appeal.

■

SELECTING AN ARCHITECT AND CONTRACTOR

Professional architects and contractors have access to and knowledge of many, many more products and materials than can be found at your local home improvement store. In addition, they are familiar with the entire building process and can advise you on all the different types of materials that you will need to select. Using the resources and skills of these professionals can greatly simplify the task of designing and building your healthful home.

Ideally, the architect and contractor should be accustomed to working with healthful materials, products and building techniques and have experience in actual construction of this type of house. They must be sensitive to the fact that building your home in a healthful manner is top priority for you and make a commitment to follow through to make sure that the house is designed and built accordingly. You may find that most professionals who are not yet experienced in healthful building practices will in fact welcome the opportunity to learn while working on your project. Look for people who are open-minded, innovative and show a personal interest.

An architect can organize your thoughts about your new home into a workable scheme. He is trained as a generalist who understands the entire building process and has practical, technical, and artistic skills. I like the saying, "The architect coordinates the art of architecture and the science of construction." It's important to choose a licensed professional architect who can take your dream and turn it into a practical working plan. You are going to live in your house a long time; you want it to be well-built and designed.

Architectural services include analyzing what is needed for the project, creating and developing the design for your home, preparing the construction documents (which include specifying building materials, brand-name products and design techniques), and providing general administration of the construction contract. You can subtract or add to the architect's role depending on what you need from him — he can be involved as much or as little as you want him to be.

A good sense of design that is compatible with your needs is also important in choosing an architect.

I like to work with people who want to have direct input into the creation and development of the design. I like to design with the owners over my shoulder, as a team, so they end up with the design they envision.

The architect also assists in awarding and executing the construction contract, so your architect will play an important role in choosing the contractor and supervising his work. The architect must be able to examine the qualifications of the contractor to determine whether the contractor can do this special kind of work.

The contractor performs the actual work of building your house. Contractors generally like to use materials they are familiar with, and have favorite places and people they purchase their materials and products from. So it's important that the contractor agree to use only the materials and products you specify and be willing to go out of his way if necessary to obtain these products and learn proper procedures for installing them. In addition, he must hire and supervise subcontractors who are also willing to work with these materials and products.

The most important person, though, on your design/build team is *YOU*. You are the *producer* (to borrow a term from the movies) of your project and are ultimately responsible for making sure everything happens according to your plan. Allow plenty of preparation time for yourself in planning the design and choosing materials and products with your architect. It is very difficult to stop construction in the middle of a project to change your mind or research a product.

Be very clear with your architect and contractor that you are in charge. Let them know you expect to have total control over all materials and methods used, that you will be making inspections, and that your approval is required before final payment. Having a cooperative and conscientious team work-working with you is worth the extra money it may cost to hire the right people.

■

BEFORE YOU BUILD OR BUY

SELECTING A BUILDING SITE

The first aspect of your healthful home that you will need to consider is the location of the site on which you will build.

Clean outdoor air is the top priority for a healthful building site. The main things to check for are:

❒ **Proximity to pollution sources** — AVOID industrial areas with heavy chemical emissions, congested urban locations, freeways or major streets, gas stations, auto body shops, or other businesses where toxic substances are regularly used, toxic dumps and sink holes, agricultural areas, underground utilities, electromagnetic radiation (locations near high-power electric transmission lines, electric generating plants, large broadcasting towers, airport radar paths, microwave relay towers), and nuclear generators.

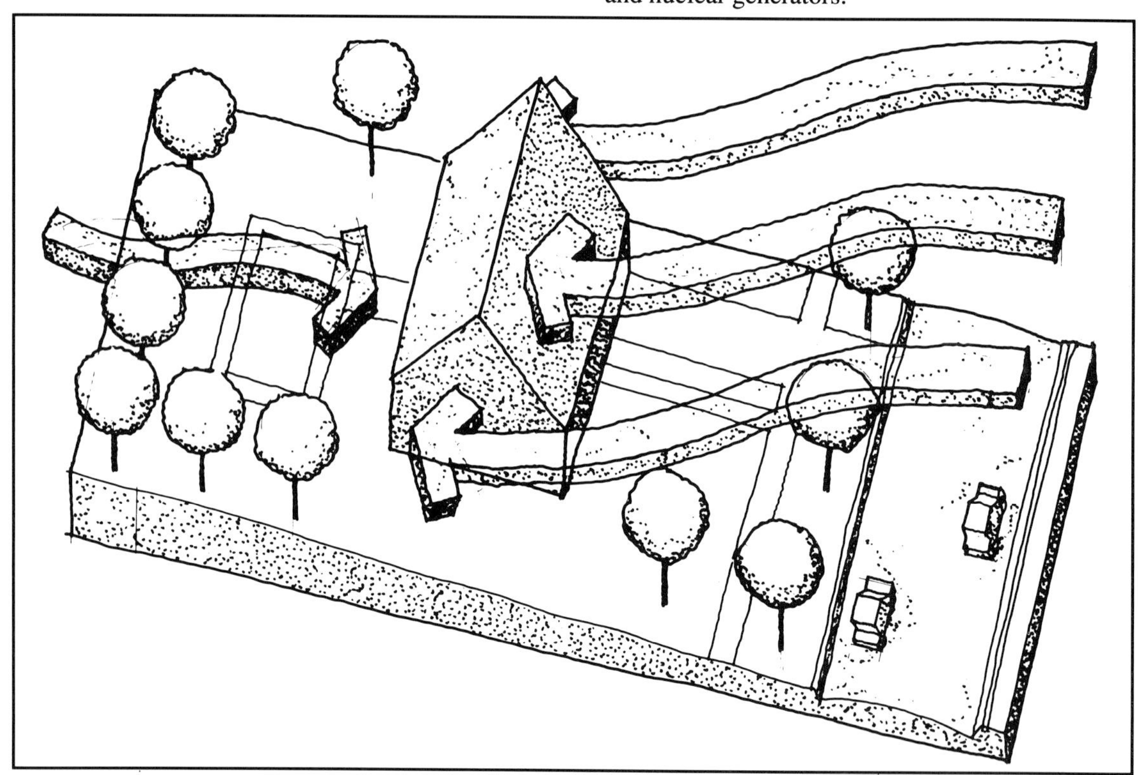

FIGURE 1: Local pollution from automobiles.

❒ **Local topography** — Certain topographic configurations are not conducive to regular air movement, causing air stagnation and build-up of pollution. Denver, Colorado, is a case in point. Located in a seemingly idyllic spot, a valley skirted by mountains, the generated pollutants from automobiles and woodburning stoves become trapped and are not allowed to be dispersed beyond the "basin". Sunlight alters the chemistry of these pollutants to create very unhealthful conditions. This same type of topographical configuration produces smog in Los Angeles.

❒ **Local weather patterns** also affect outdoor air quality. Best are windy, rainy areas, with prevailing winds in a favorable direction.

The recommended optimum healthful site would be a rural setting on the side of a metropolitan area opposite the direction of the prevailing winds (you can find out wind directions from the local air quality agency), preferably away from farming belts. Since wind speed affects the amount of air infiltrating the structure, it would be good to locate in windy areas.

Next, evaluate the neighborhood:

❒ **Characteristics :** Are the houses well spaced? Is there plenty of land in each lot? Is it zoned for dense residential development? Are the other houses in the neighborhood of good quality?

❒ **Future planning :** Are there current plans for or the likelihood of future industrial or research and development buildings, airports, highways, or main streets? Check with your local planning and/or zoning department for information.

❒ **Proximity to shopping, recreation, school, work :** You don't want to be too close because of pollutant concentrations, but you don't want to be so far away that you have to drive many miles just to obtain basic necessities such as food and other services.

Be sure to discuss the following points with your architect, who can advise you on the suitability of your building site:

❐ **What is the geologic composition of the soil?** Radon is the prevalent problem in the soils of certain areas of our country (see page 11) but asbestos is a close contender. Asbestos in the soil will probably be the next area of concern as it is now in our interior environment. Also look for well-drained locations where soils are suitable for construction.

❐ **Is the lot big enough to build on and is there enough buildable land?** For health purposes, it's best to have plenty of space around you to act as a buffer against possible toxic practices of your neighbors. Some land is unsuitable for building at all, and your architect can help determine this.

❐ **Are there objectionable smoke, odors, noises, unsightly features, or similar nuisances?**

❐ **Is there vegetation that can be beneficially integrated into the site design?** Surrounding trees and vegetation deflect winds, clean air and facilitate climate control for the house. Planting of bushes and shrubs attracts wildlife and architecturally integrates the building and the natural site.

❐ **Is there adequate sunlight at all times of the year?** Are there many trees blocking sun needed for light and heat? Does the angle of the winter sun place it behind a nearby hillside? Is the site located in a winter cloud belt?

❐ **What is the direction of the prevailing winds?** Is the lot upwind of the nearby street to minimize traffic fumes? Do the prevailing winds bring fresh or polluted air from nearby areas?

❐ **Can the house be oriented well?** Will the lot allow orientation of the house living areas to the south or southeast to maximize passive solar gain? Will windows have pleasant vistas and can they be oriented to take full advantage of breezes for ventilation?

❐ **Does the site have nearby natural resources?** Is there a small stream to take advantage of miniature hydro-electric generating capability, or is it a windy site that can incorporate a wind generation system? Does it have conditions such as a slope where solar photovoltaic electricity can be generated?

BUYING AN EXISTING HOUSE

If you are considering purchasing an existing house, check out the building for the following:

❒ **Source of energy** — Watch out for homes with gas heating and appliances. Because of the health effects of combustion byproducts, these homes may require changing the heating system to a less polluting energy source such as electricity or solar. If you want to continue to use natural gas for economic reasons, you may want to put the furnace in a sealed or remote location, such as a shed or a sealed-off part of the house with an outside entry only, and use the furnace to heat a circulating hot water heating system.

❒ **Pesticide history** — Find out as much as you can about what types of pesticide treatment have been applied to the house, and have an expert check for past pest damage, which will also give you clues. Steer clear of homes that have been treated with chlorinated pesticides such as heptachlor-chlordane. These pesticides are especially toxic and long lasting, and can contaminate a home for many years, especially if they were originally applied incorrectly.

❒ **Flooring** — Look for homes with existing hardwood, ceramic tile, brick, stone, concrete, terrazzo or old, hard vinyl tile floors. If floors are carpeted, pull up a corner to see what's underneath. Homes that pre-date the mid-1960's probably have wood flooring underneath the carpets that can be refinished if necessary. In more recent homes, the subfloor might be concrete, plywood, or particleboard, all of which will need new finish flooring. Plywood and particleboard subfloors should in addition be sealed to reduce formaldehyde emissions.

❒ **Built-in cabinetry** — Check kitchen and bathroom cabinets, wood paneling, shelving, and the undersides of plastic laminate countertops for exposed particleboard and plywood that may be emitting formaldehyde. These wood products may be particleboard covered with a wood veneer, so check corners, interior shelves, shelf bracket holes, and undersides for visible particleboard.

POLLUTION TEST KITS

The following companies provide pollution test kits that can aid your evaluation of the safety of air quality in an existing house:

National Mine Service Co.
US Route 22-30
West Oaksdale PA 15071
"Gasbadge" monitoring of organic gases and vapors by adsorption on activated carbon followed by gas chromatographic analysis. $26 per badge plus minimum $24 for analysis.

Du Pont Company
Applied Technology Division
Concord Plaza
Clayton Building
Wilmington DE 19898
"Pro-Tek Badge" monitors formaldehyde by adsorption followed by spectrographic analysis. $18 per badge, $80 for analysis.

3M Company
Occupational Health and Safety Products Division
Building 220-7W
St. Paul MN 55144
"Formaldehyde Monitor 3750" monitors formaldehyde by adsorption followed by spectrographic analysis. $35 per monitor includes analysis.

Bendix Corporation
12345 Sparkey Road
Largo FL 33543
"Gastec Detector Tube" provides grab sample monitoring for formaldehyde, carbon monoxide, and nitrogen dioxide in less than one minute by chemical reaction in detector tube. $130 for pump plus $2 per detector tube.

Lab Safety Supply Company
PO Box 1368
Janesville WI 53547
"Personal Protection Badge" provides integrated passive monitoring of carbon monoxide by abdsorption and color change. $3 per badge.

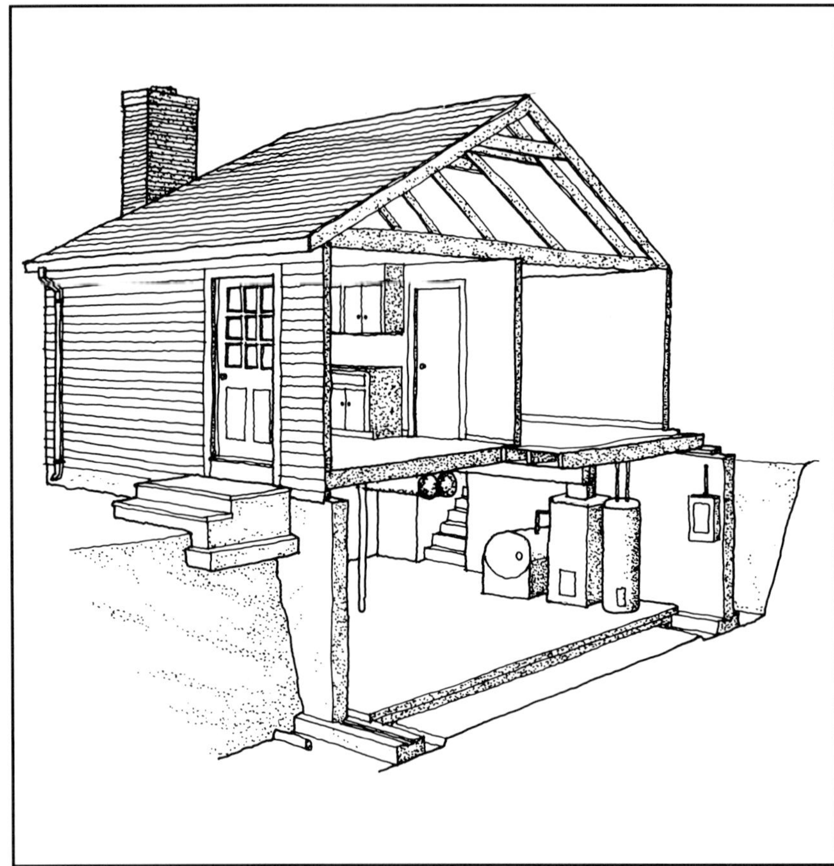

FIGURE 2:
Where asbestos might be found.
Look for:
- Ceilings with a "cottage cheese" look
- Duct work and steam pipes covered with asbestos insulation
- Vinyl floor tiles that are badly cracked or sanded
- Patching compounds or textured paints

❒ **Insulation** — Check the type of insulation used and avoid entirely any house that has been insulated with urea-formaldehyde foam insulation (UFFI), now banned due to formaldehyde hazard. Check also to see if UFFI has been removed. It appears as a light brown, foamy mass, or as a residual crust or powder. Look for evidence of UFFI around attic and basement perimeters, especially where wires or pipes enter walls, and beneath electrical cover plates on outside walls.

❒ **Garage** — If you plan to put your car in the garage, it would be preferable to choose a house with a detached garage, as car exhaust fumes can enter the living areas from attached garages. If you are "attached," however, to a house with an attached garage, I've recently designed a mechanical ventilation system that will automatically ventilate the garage when pollutant levels are high.

❒ **Asbestos** — In houses built between 1920 and 1978, asbestos products may have been used. There may be no danger from asbestos products if they have not been damaged in some way that would release asbestos fibers, and some installations can be sealed against future problems. Asbestos removal, if it becomes necessary, must be done by a contractor approved by a local governing agency. This can be very costly, so find out if this might be a problem before you buy.

RADON

Of all the possible household pollutants, radon is considered the most hazardous. A radioactive gas that is virtually nondetectable without professional testing, radon comes from the natural breakdown of uranium. It occurs in concentrations that vary widely from area to area, even from neighborhood to neighborhood. Seeping from the earth, it becomes diluted in the outdoor air in low, negligible concentrations. Accumulating in an enclosed house, however, levels can rise to concentrations high enough to cause lung cancer. The level of risk depends on the concentration of radon and the length of time you are exposed.

Most homes are not likely to have a radon problem; however, a significant number have low to moderate levels which over a long period of exposure increase cancer risk. A small percentage of homes on testing show a high concentration of radon, which calls for immediate action to lower the exposure and consequent risk involved. Since it is next to impossible to predict which areas or even which houses in a given area are contaminated, it is very important to have your home tested using a simple detector that is put in place for a given length of time. You can get radon detectors from private firms or from your state or local government.

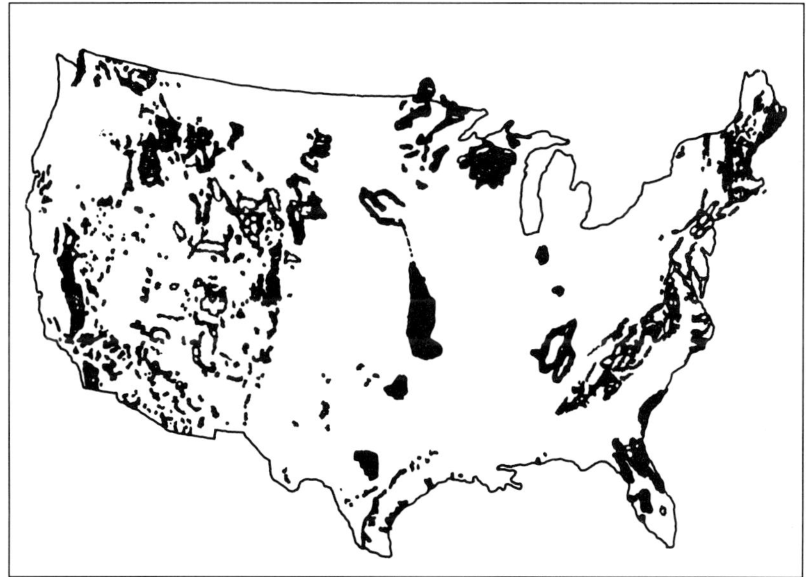

FIGURE 3: Areas with potentially high radon levels.
This map, produced by the U.S. Environmental Protection Agency, shows geographical areas where there are radon-producing earth formations which include granite, phosphate, shale, and uranium. Some concrete, brick and gypsum board may have significant radon levels if made from natural materials taken from high-radon locations. If large portions of the house are stone, brick, or concrete, it should be tested for radon.

Upon receiving the results and an interpretation of the actual risk you face, there are steps of varying difficulty, cost, and effectiveness that should be taken, with immediacy depending on concentration. Most remedies require the skill of a professional experienced in radon reduction procedures and design. The type of reduction measures that will be necessary depend upon the unique characteristics of the house, and often a combination of methods will be necessary to achieve acceptable results.

The most obvious, immediate, and effective step is to ventilate your house, replacing the radon-laden air with outdoor air, either by opening windows or with forced ventilation. Ventilation is only effective up to a certain level of concentration, however.

More permanent remedies include providing appliances — such as furnace, clothes dryer, and fireplace — with external air sources, so that radon is not drawn in by the lowering of indoor air pressure caused by indoor air being vented outdoors. Another necessary step, either alone or in combination with others, is to cover exposed earth in basements, crawl spaces, drain areas, sumps, etc., and to seal cracks and openings in concrete floors and walls, around pipes, at wall joints, etc. Some of these areas might be accessible only at considerable expense, such as spaces concealed by brick veneer or fireplaces.

More drastic measures may need to be employed in cases of high radon levels or if other methods prove ineffective. It may be necessary to install a continuous circuit of drain tiles around the foundation, draining away water and thereby venting the radon. Another approach to venting radon is to insert venting pipes through the concrete foundation slab. A major source of radon seepage might be through hollow blocks forming basement walls. If sealing the top holes of the wall is not effective, it may be necessary to install a block-wall ventilation system which provides either wall suction or wall pressurization (blowing air into the space to prevent intake of radon-laden air). This is a costly but highly effective method.

More information on all aspects of radon measurement and reduction can be obtained from your state radon protection office or from your EPA regional office.

The bottom line is: before you buy a house or a lot, get it tested for radon. Better to know you're safe than have the difficult and expensive problem of radon control later on.

RADON TEST KITS

Air Chek, Inc.
180 Glenn Bridge Road
Box 2000
Arden NC 28703
800/AIR-CHEK or 800/25-RADON

Terradex Corporation
3 Science Road
Glenwood IL 60425-1579
800/528-8327, in Illinois 312/755-7911

The Radon Project
University of Pittsburgh
Pittsburgh PA 15260
Profits from sales support radon research.

Environmental Measurements Lab
U.S. Department of Energy
376 Hudson Street
New York NY 10014
Approximately $20 each

■

PLANNING YOUR HOME

Architects are experts in helping you define and organize an overall description of the features you want in your home: the number and size of rooms, their functions and relationship to each other, traffic patterns, service and utility requirements, engineering requirements (electrical, mechanical, structural), finishes, colors, healthful materials, automobile parking, and landscape requirements.

When designing a healthful house, there is more to consider than aesthetic appeal. If cost is a factor, an efficient floor plan can reduce overall building costs and balance out the higher cost of some of the interior finishing materials.

ENERGY EFFICIENCY

Weatherstripping, caulking, and sealing a house for energy efficiency is a good method of minimizing winter heat loss, but this "tightening" also traps pollutants indoors. Leave a few cracks to allow the house to "breathe" and exchange indoor air for fresh outdoor air at a slow but regular rate.

Winter heat loss can be minimized by using effective insulation in walls and floors in addition to ceilings. Prevent heat loss through windows by using double-pane glass, storm windows, shades or heavy drapes made of a natural fiber, or external shutters.

Keeping your house cool in the summer can be accomplished with good house design and integrated plantings. Position and design your house to take advantage of prevailing summer breezes for thorough ventilation. Design adequate overhangs or trellises to the south and louvers or shades to the west to block hot summer sun. Orient windows so as to flood rooms with light in the winter, yet be shaded as the sun passes overhead in the summer. Deciduous trees and large shrubs to the south and west will provide cooling shade just when you need it, yet will shed their leaves and allow warming sunlight during the winter. Deciduous vines on trellises will do the same.

Houses should have attics to provide a layer of protection from solar heat, since most summer sun is directly overhead. Proper attic insulation and ventilation, such as continuous ridge, gable, and soffit vents, will help dissipate attic heat build-up, as will an attic fan.

DAYLIGHTING

Filling your home with natural light is pleasant, healthful, and economical. Design plenty of windows and skylights into your house, and make sure they are oriented well to receive useful natural light, yet be protected in the bright summer months when too much sunlight can cause glare, profuse heat, and damage from excess ultraviolet light. You may also want to design your house to take advantage of sunlight to provide power for your home.

When designing your floor plan, keep in mind the type of natural light you will want for each room as well as the views you will see out of windows. Regardless of the orientation of a room, you can add more light with many large windows, skylights, greenhouse windows, or walls of glass blocks, and decrease the amount of light by installing only a few small windows or adding awnings.

Northern exposures are the darkest and coolest, having the most consistent light because the sun never shines directly through north-facing windows.

Eastern exposures provide cheerful morning sun. You might prefer this exposure for a kitchen or breakfast room, or a bedroom if you like to wake up with the sun.

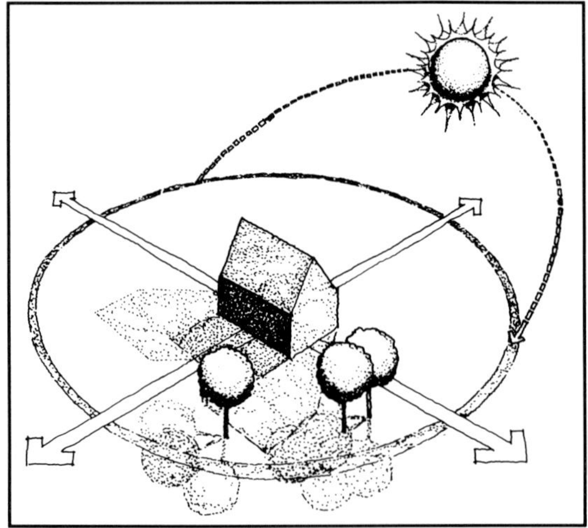

FIGURE 4: Orientation of house.

Southern exposures receive the most sun in the wintertime when the low-angled rays shine directly in. Many south-facing windows can provide significant wintertime warmth and are the preferred exposure for living areas, sunrooms and greenhouses.

If at all possible, avoid placing windows on the west to prevent overheating rooms in the summertime.

VENTILATION

Proper house site location and shape are necessary in optimizing natural ventilation. The advantages of natural ventilation are that it is available at no cost and it is relatively quiet (except for an occasional bird song). The disadvantages include lack of control and the introduction of airborne particulates such as pollens, dust and vehicle pollution.

In most regions prevailing winds come primarily from one or two directions. You can find out about winds from the local air quality management agency or weather bureau. Place windows strategically to maximize cross-ventilation. This includes placing them on the back side of the house since local automobile pollution decreases rapidly as you move away from the street.

The shape of the house is also important. A more linear and narrow house will allow for better cross-ventilation. This has the added bonus of allowing more opportunities for passive solar design.

Further information on house ventilation can be found in Part Two, Division 15: Mechanical.

ENVIRONMENTAL INTEGRITY AND LANDSCAPING

When designing your healthful house, you must also consider the surrounding environment and the possible health effects of products that will be used in the garden areas.

The simplest way to minimize use of herbicides and pesticides is to maintain the natural vegetation as much as is practical. Consider the placement of existing trees and shrubs and design the house in such a way as to harmoniously blend into the existing ecosystem.

Plan to maintain landscaping without the use of chemical weed and bug killers or chemical fertilizers. Vegetable gardens can flourish when tended with organic methods, as will decorative plantings.

TERMITE CONTROL

Termites are generally controlled by the use of toxic chemicals. A healthful house is protected from termite invasion through design features which consider the habits of the termites.

The time to make a house termite-proof is when it is being built — don't skimp on the termite protection and then have to use toxic chemicals to alleviate the problems later.

Like any other living creature, termites need the proper environmental conditions in order to live. They love the dark, need wood or other cellulose materials to eat, and a considerable amount of moisture. Without these conditions, termites die.

Subterranean termites, which of 58 species of termites in the United States are the ones who do 99% of the damage, build their nests or colonies either in the ground or in wood that is in contact with the ground. They can enter houses by flying through ventilators and other openings — under the house, through unfinished attics, all they need is a tiny crack. They can also work their way up from the earth through a crack in the foundation, or by building tunnels up the sides of foundation walls or pipes.

Installation of noncorrosive metal termite shields at basements, crawlspaces, or chimney foundations; on concrete piers with saddle and wood post; and at entrances with wood steps or floor is the simplest and most effective protection. When properly installed, these shields prevent termites from reaching the woodwork from the earth. Copper shields are preferable, but galvanized steel is acceptable and more cost-effective.

In addition to the use of shields, the following precautions should be taken when designing and building your house:

❒ Design and construct foundation walls and supporting piers to be free from future cracks or fissures through which termites can attack.

❒ All wood in contact with the ground should be heart tidewater red cypress or heart redwood or treated with a natural wood preservative.

❒ Provide sufficient cross-ventilation for unexcavated areas under the house.

❒ Screen all openings for ventilation under the house or in the attic with 18-mesh metal screening.

❒ Remove all wood scraps and stumps from the soil surrounding the house.

❒ Wooden floor joists should be at least 24" above ground and wooden construction on the exterior should be at least 6" above ground.

The *Common Sense Pest Control Quarterly* (published by the Bio-Integral Resource Center, PO Box 7414, Berkeley CA 94707, 415/524-2567) has excellent literature on termites and how to treat them using the least toxic methods.

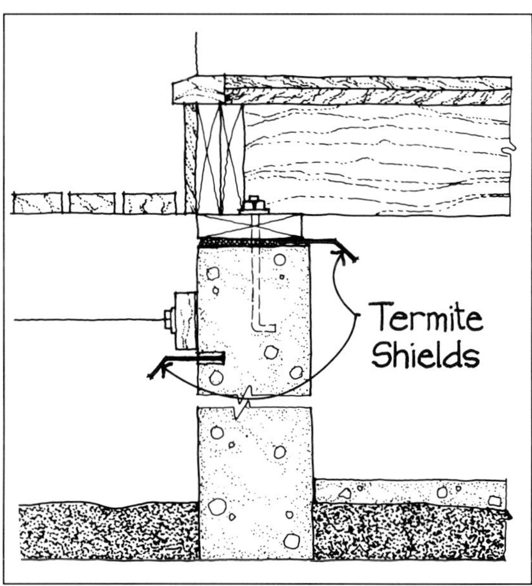

FIGURE 5: Section showing termite shields.

BUILDING AND FINISHING MATERIALS

Choosing proper materials is the most important factor in building a healthful home, and it will be necessary to consider these materials in the design phase. Healthful materials, products, and installation details will be presented in "specifications format" in Part Two, but for your design reference, here's a chart of materials that can be used:

BUILDING MATERIALS

FOUNDATION	Concrete Glazed brick Concrete block Stone Brick
FOUNDATION COATINGS	Cement-based sealants Specially formulated low-toxicity coatings
BUILDING FRAME	Reinforced concrete Masonry Brick Concrete block Stone Lightweight steel Lumber Wood-metal prefabricated systems
EXTERIOR SIDING	Brick Porcelain-enameled metal siding Stucco on metal lath Wood
ROOFING	Steel — galvanized or factory-applied baked-on enamel Aluminum Tiles — cement, terra cotta Wood shingles Terne metal Copper Slate
INSULATION	Cementitious magnesium-oxide foam Aluminum Perlite
VAPOR BARRIER	Aluminum foil Foil-backed paper
INTERIOR WALL AND CEILING COVERING	Plaster on metal or wood lath Brick with mortar joints Solid wood paneling (maple, pine, cedar, birch, alder, poplar, spruce, fir, etc.) Gypsum board, taped and spackled with a nontoxic joint compound
SUBFLOOR	Pre-cast concrete Solid wood plank over structure
UNDERLAYMENT	Cellulose-type material
DOOR AND WINDOW FRAMES	Metal Porcelain steel Enameled metal Solid wood
PLUMBING	Water supply: copper pipe with no-lead solder joints Drain: copper and cast iron Electric hot water heating
WIRING	Metallic-clad wiring Metal receptacles
HEATING/ VENTILATION/ AIR CONDITIONING	Radiant (oil- or water-filled radiators and baseboard units) Electric heat pump Electric central furnace Air-to-air exchanger (heat recovery ventilator)

	Ducts: sheet metal (cleaned)
APPLIANCES	Electric range, clothes dryer Porcelain enamel interior refrigerator Induction range Water filtration equipment Air filtration equipment
ADHESIVES	White glue Yellow glue Mortar adhesives Water-based carpet adhesive Water-based lino-leum adhesive
SEALANTS	Clear silicone Nontoxic caulking
WEATHER-STRIPPING	Felt strip bound with metal Hard vinyl strip
TERMITE CONTROL	Metal termite shields

FINISHING MATERIALS

DOORS	Wood panel Metal
WINDOWS	Metal Wood
WINDOW TREATMENT	Natural fiber draperies or curtains Woven wood shades Wooden blinds Mini blinds Grasscloth shades Natural fiber fabric shades Rice paper shades
PAINT	Specially formulated low-toxicity paints Natural-ingredient paints Water-based latex paint (cured)
WOOD FINISH	Specially formulated low-toxicity wood finishes Natural-ingredient wood finishes
WALLPAPER	Plain paper wallpaper Natural wallpaper paste
FLOOR COVERINGS	Ceramic tile Solid hardwood (oak, maple, beech, pine, ash) — planks or tiles Natural linoleum Marble and other stone tiles Terrazzo Stone Brick Natural fiber carpet
CABINETS	Solid wood Metal
COUNTERTOPS	Ceramic tile Hard plastic lami-nate (Formica) Marble or granite slab Corian
SHOWER STALLS	Ceramic tile Pre-cast terrazzo
PLUMBING FIXTURES	Stainless steel Porcelain Enameled steel Enameled cast iron
ELECTRICAL COVER PLATES	Steel Ceramic Enameled Brass
LAMPS AND LIGHT BULBS	Metal and glass fixtures Incandescent bulbs

MAKING YOUR HEALTHFUL HOUSE COST-EFFECTIVE

Designing an efficient floor plan is a key factor in building a cost-effective healthful house because it can reduce overall costs and make up for the higher cost of some healthful materials. Here are some tips for efficient floor plans:

❒ Maintain an efficient layout (see the prototype healthful house layouts at the end of this book) where spaces flow from one to the other, the plumbing is organized, and there is a main hallway connecting the rooms.

❒ Vertical multiple-story designs are less expensive to build than the same square footage on one story, and it saves destroying the natural habitat too.

❒ Rectangular designs have fewer changes in the directions of the walls of the exterior envelope and are simpler and less expensive to construct.

❒ Contemporary designs with open plans have fewer interior walls to construct (and have the added benefit of allowing better circulation of air).

❒ Symmetrical and repetitive details are more cost effective — the more repetitive the task, the faster the tradesmen can work, thereby saving labor cost.

❒ Design space efficiently and build a quality house with less square footage. Remember the old adage, "Less is more".

N O T E S

■

THE CONSTRUCTION DOCUMENTS

Accurate construction documents are vital to the proper construction of your healthful home for they are the legally binding written agreements you have with the architect and contractor regarding the products they will use and the methods, style and quality of construction. There are three basic parts:

❒ *The Agreements and the General Conditions of the Contract for Construction* are separate documents which are used together to outline the legal agreements between you and the architect and you and the contractor.

❒ *The Contract Drawings* indicate the physical design: floorplans, elevations, building sections, and details.

❒ *The Specifications* describe what materials and products are to be used, how they are to function and where they can be obtained.

Each part of the construction documents can be used to insure that a healthful house will be properly built.

THE AGREEMENTS AND THE GENERAL CONDITIONS OF THE CONTRACT FOR CONSTRUCTION

The American Institute of Architects (AIA) provides standard forms for the Agreements and the General Conditions of the Contract for Construction documents, available from all architects' offices.

"The General Conditions of the Contract for Construction" (AIA Document A201) are just that, general conditions which apply to all contracts. They are used in conjunction with "Standard Form of Agreement Between Owner and Architect" (AIA Document B141) and "Standard Form of Agreement Between Owner and Contractor" (AIA Document A101).

Since the architect is responsible for specifying products and building techniques to the contractor, in your Agreement with him add a paragraph at the end under Article 15, "Other Conditions or Services", such as the following:

"The Architect shall incorporate details on the plans which reflect healthful building techniques and specify only those brand name building materials and products which are approved by the Owner as meeting his requirements for a healthful building."

Since the contractor is responsible for implementing building techniques and using the products specified by the architect, in your Agreement with him add a paragraph at the end under Article 7, "Miscellaneous Provisions", such as the following:

"The Contractor shall implement the contract documents as indicated, using only healthful building construction practices and specified products. Substitutions can only be made with the approval of the Owner."

THE CONTRACT DRAWINGS

The Contract Drawings are the "blueprints"—the pictures that show the contractor exactly how to build and what the finished product should look like.

When building a healthful house, it is important that these drawings are complete and clear, for to avoid using unhealthy products you will be using some building techniques that some contractors are not familiar with. These techniques will be discussed in detail later on in this book.

THE SPECIFICATIONS

The Specifications are the most important part of the construction documents for a healthful house, for this is where the architect specifies the materials and products that will be used to build your home, and how they will be installed. Standard products and building techniques are often used if none are specified, and each architect and contractor has his own preferred methods and materials. So it is imperative that your specifications be well written and exact.

All product research and selection must be done in the planning stage with your architect and written into the specifications, for clarity and your legal protection. *Don't expect to be able to go directly to the contractor during construction and tell him which products to use. Write all product selections clearly in the specifications.*

The specifications are the most difficult part of planning for the architect because he is required to identify manufactured products by trade name or brand name and control substitutions. Generally, this is handled in one of five ways:

❐ **Product Approval Specifications** list one or more trade names for each product. Requests for substitutions must be made a given number of days prior to bid opening, and all approvals are issued as a written addendum to the construction documents. These specifications are the best to use in most cases because they limit the products to be used but also provide for the submission of desirable products by the contractor that the architect or owner may not be aware of.

❐ **Contractor's Option Specifications** list every trade name which is acceptable for each product. The contractor may at his discretion use any of the products listed without consulting the architect or owner, but can not substitute any products which are not listed. These specifications give greater freedom of choice to the contractor and might result in greater compliance to the project requirements.

❐ **Non-Competitive or Closed Specifications** list only one trade name for each product required. The contractor must provide and install this product without substitution. These specifications should be used if only a narrow range of products are acceptable to you. The disadvantage is that it limits the contractor's options of competitive pricing and preference for workability of products.

❐ **"Or Approved Equal" Specifications** list one or more trade names for each product and follow each with the phrase "or approved equal". This allows the contractor to request permission from the architect to substitute an unlisted product the contractor considers to be equal. These specifications are not recommended when specifying healthful products because it opens the possibility for unacceptable products to be used without the owner's approval.

❐ **Product Description Specifications** describe completely all details, qualities, functions, and sizes of a product without mentioning a trade or brand name. Any product which meets all of the detailed specifications will be approved. These specifications are not recommended when specifying healthful products since they don't give the contractor information about the brand-name products that exist.

Specifications are written in a standard industry format to make them easy to use by all architects and contractors. Your architect will probably write the specifications for your house according to the *Masterformat Manual of Practice* issued by the Construction Specifications Institute, so I'll be using this format in the following pages to display and discuss the many brand-name products, materials, and special building techniques that are available for you to use in designing your healthful house.

N O T E S

■

PART TWO

Specifications for Healthful Houses

The purpose of these specifications is to describe materials and products that can be used in the building of healthful houses. Not all sections of the standard specifications are included, only those which are relevant.

Healthful materials of various sorts are described in general terms in the specifications when any generic brand is acceptable. Often there are many manufacturers who make similar products and have limited regional distribution. Your architect has access to information on local sources for many of the materials mentioned in his *Masterguide: The Official Specifying and Buying Directory of the AIA.*

Brand name products, along with purchasing information, are mentioned when it is important to specify a particular product by its brand name. These are products I personally have experience with and can recommend. The names and addresses of manufacturers are given so you can contact them to find out the local distributor of the products mentioned. In most cases, you will not be able to purchase directly from the manufacturer.

The materials and products used to build your house are arranged in the specifications according to the industry standard as outlined in the *Masterformat Manual of Practice* issued by the Construction Specifications Institute. If you are unaccustomed to reading specifications, this order may seem rather strange at first, so here is an alphabetical index to products you may be looking for:

INDEX

N O T E

There are three companies whose products are mentioned repeatedly in these specifications. Rather than take up space repeating the general philosophy behind these products and the manufacturers' addresses over and over, I'll cover them here:

AFM Enterprises, Inc. (1140 Stacy Court, Riverside CA 92507, 714/781-6860) has a line of finishes, paints, sealers and cleaners which are low-toxicity synthetic formulations that do not contain petrochemicals. While the company is unwilling to divulge the exact ingredients of these products, experience in working with these products has shown that they are quick-drying, low-odor products that are acceptable even to those who have individual sensitivities to many petrochemical products.

Auro Products, imported from West Germany by **Sinan Company** (PO Box 181, Suisun City CA 94585, 707/427-2325), is a line of paints, finishes, adhesives and cleansers made from 100% natural ingredients which are compatible with our natural world as well as being safe for our personal health.

The Livos Plantchemistry (614 Agua Fria Street, Santa Fe NM 87501, 505/988-9111) line of paints, finishes, adhesives, and cleaners are also made from 100% natural ingredients, and are safe for our health and our planet.

01020 ALLOWANCES

In his bid for the entire project, the contractor allots a specified amount of money for purchase of products and materials.

You may wish to take the responsibility yourself for choosing and purchasing items such as kitchen and bathroom cabinets, appliances, countertops and other accessories, in which case an "allowance" of a specified amount of money for each item can be written into the specifications for you to use for this purpose.

01030 SPECIAL PROJECT PROCEDURES

It is of utmost importance that you keep in focus the fact that the purpose of this project is to build a healthful house. You can use this section to make statements which give more information on general policies to be observed throughout the project.

PRODUCT SUBSTITUTION

First and foremost is to state that *specified healthful products must be used in all cases*. There is to be no substitution of products or materials or use of unapproved products or materials that may be needed but are not mentioned in the specifications. Use specified products only or get approval first!

FORBIDDEN PRACTICES

You might also want to list specific unhealthful practices that you forbid, such as:

❒ *"**No smoking** allowed during interior construction."*

❒ *"The use of any toxic substances such as **pesticides** or **noxious cleaning products** is **prohibited** anywhere on the site."*

❒ ***"Do not use gasoline-powered generators** inside the building."* (Generators are used to provide electricity for construction machines — do not use indoors to prevent absorption of gasoline by materials if accidentally spilled.)

DIVISION 1

CONTRACT REQUIREMENTS

GENERAL CLEANUP

General cleanup at the end of the project will be done with ordinary cleaning products unless you specify otherwise.

Most cleanup can be done with plain water — damp rags can be used for dusting, for example, as effectively as an aerosol product. Clean mirrors and windows with a solution of half distilled white vinegar and half water (adding more vinegar if necessary for strength) and crumpled newspapers. Have the workers protect materials of large surface areas (such as floors or walls) as recommended by the manufacturer of the finish to eliminate the necessity for extensive cleaning at the close of construction.

If you have your own favorite methods or cleaning solutions, write them clearly in the specifications under this section. You can help your workers comply by providing your preferred healthful products (your local natural food store will have a good selection) and being at the site at the beginning of cleanup to *show* them exactly how you want it done.

Brand name products:

Bon Ami Polishing Cleanser (Faultless Starch/Bon Ami Corporation, Kansas City MO). A nonchlorinated scouring powder good for cleaning sinks, ceramic tile, bathroom and kitchen fixtures. Available at all supermarkets and hardware stores.

AFM Super Clean (see note on page 28). An excellent all-purpose cleaner.

AFM X158 Mildew Control (see note on page 28). Can be applied to mold-prone areas (bathrooms, under kitchen sinks, windowsills, etc.) as a prophylactic treatment to prevent the growth of mold on surfaces. To remove mold, use AFM Safety Clean, an anti-fungal bacterial cleaner.

Auro No. 411 Organic Soap (see note on page 28). Pure potassium soap made of plant oils and beeswax.

Livos Avi Soap Concentrate and Livos Latis Natural Soap (see note on page 28). For floors, tiles, and window frames.

01040 COORDINATION

The contractor is responsible for coordinating all the tradesmen on the project, supervising them and making sure their work meets the standard. You can reinforce your position regarding the healthfulness of this project by inserting a statement in this section such as:

"The Contractor shall coordinate the tradesmen and provide the necessary training to properly install healthful products and employ healthful building techniques."

01400 QUALITY CONTROL

Of course, you will want your healthful home built to the same standard of quality as it would be if other, more familiar materials were being used. Some contractors, however, may be unwilling to guarantee the quality of their finished work using healthful products that they are unfamiliar with. **Do not** agree to this. If they follow the manufacturer's instructions, they should have no problem maintaining a standard of excellence.

To make this expectation legally clear, insert a sentence in this section such as:

"The Contractor shall maintain, with the installation of healthful procedures, the same standard of quality expected in standard building practices."

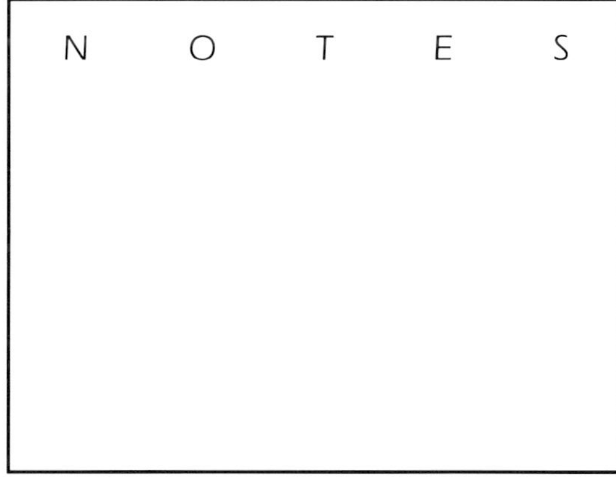

■

DIVISION 2

SITEWORK

02100 SITE PREPARATION

If you wish to maintain the natural quality of your site, you must specify so in this section. Contractors generally focus on the house they are building more than the environment in which they are building it, so if you have favorite trees and bushes that need to be preserved, state it in writing in the contract documents. Ask workers to be aware of the natural vegetation and take care to keep it intact as much as possible. (Of course, if you prefer to have your site totally cleared and leveled, that's OK too, and this is the section to state that.)

Also specify in this section that the site is to be prepared in such a way that allows for the house to be oriented according to the plan, to insure that the building is placed properly on the site.

02200 EARTHWORK

SOIL TREATMENT

Do not treat soil with any sort of chemical treatment. See page 16 for controlling termites without soil treatment.

02500 PAVEMENT

Surfaces can be healthfully paved with concrete, brick, or stone. Pea gravel with cedar timber edging anchored with reinforcing bars also makes a nice paving. Avoid asphaltic paved driveways.

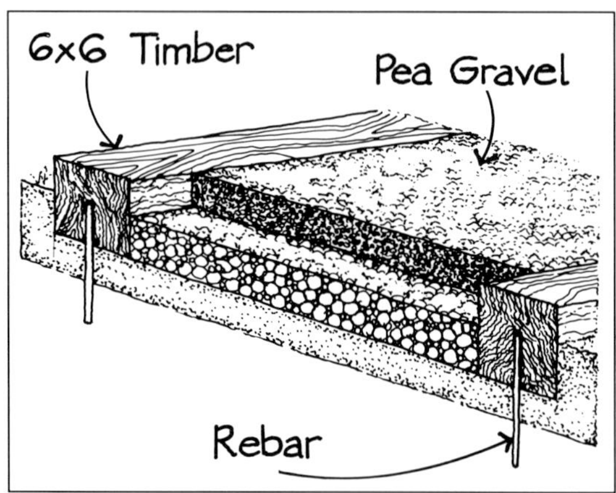

FIGURE 6: Section of walkway paving.

■

When mixing concrete, avoid admixtures (water repellents, coloring agents, additives to accelerate or retard the drying process). They are not necessary to the performance of concrete.

03100 CONCRETE FORMS

Concrete forms are used for concrete foundation walls and footings, elevated floor slabs, and exterior walls in concrete houses.

In wood form work, the forms are frequently coated to aid removal and clean-up. When the concrete dries, the coating remains.

Specify metal formwork in lieu of wood forms — no coating is used on the metal forms.

03300 PRE-CAST CONCRETE

Structural concrete — lengths of pre-cast concrete material — can be used in conjunction with concrete blocks to form an inert structure. Pre-cast concrete does not contain admixtures. These slabs can be used as floors, for a cantilevered porch or deck — be creative.

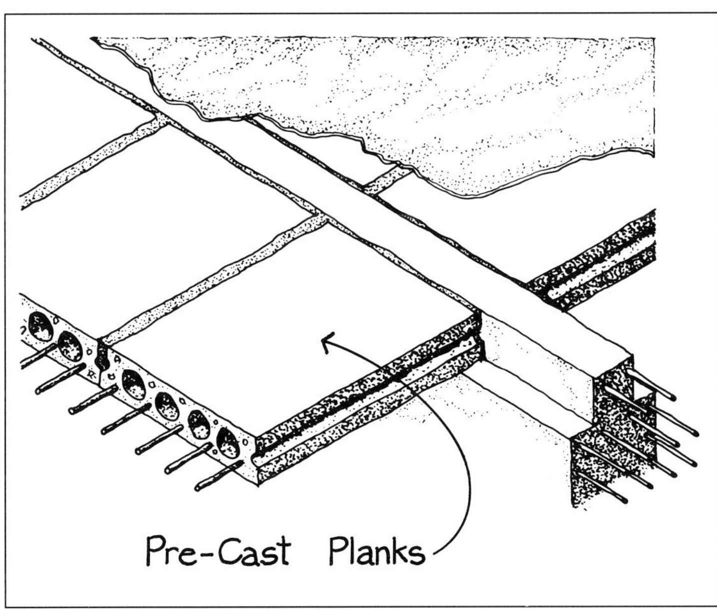

FIGURE 7: Pre-cast concrete flooring systems.

■

DIVISION 4

MASONRY

04200 UNIT MASONRY

Brick can be used as a veneer over wood frame construction or concrete block walls. Brick comes in many colors, shapes and sizes.

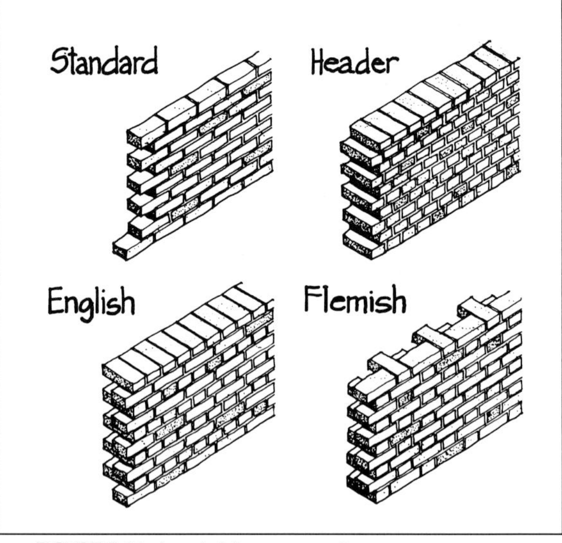

FIGURE 8: Various brick veneer patterns.

FIGURE 9: Architectural concrete blocks.

Concrete blocks come in a wide variety of shapes and sizes:

❒ **Structural glazed tile** — concrete block with a glazed or ceramic face that can be used to construct a wall, etc.

❒ **Exposed aggregate concrete block** — concrete block with cement washed off before curing to expose granular material.

❒ **Architectural blocks —** scored, hex, offset, slump, indented, sculptured, waffle, ribbed, split ribbed, fluted, splitfaced, molded face and fluted concrete block — all can be used to construct interior or exterior walls, giving a nice textured finish.

❒ **Ceramic veneer** — ceramic tile anchored to concrete block.

Terra cotta tiles, made of hard, unglazed, fired clay, can be used for ornamental work and roof and floor tile or wall facing.

Glass blocks are ideal for constructing inert walls in places where you want to let additional light in but also retain privacy.

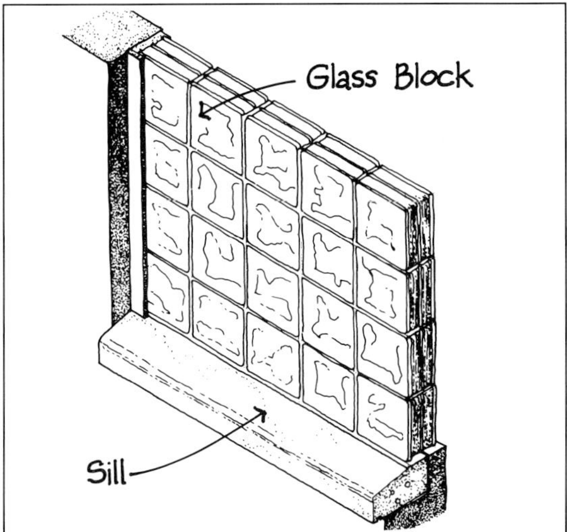

FIGURE 10: Glass block wall.

04400 STONE

There are many types of stone that can be used in construction, either as a building material, such as building a stone wall, or as a facing material, such as using marble tiles for flooring.

Inexpensive common rock is selected or processed by shaping, cutting, and sizing for building or other uses. "Rough stone" has been washed up from a riverbed or quarried and left in the various shapes, sizes and textures as it occurs in nature. "Cut stone" has been processed, cut to stock shapes and sizes.

Special higher quality, unusual stone — marble, limestone, granite, sandstone, slate — is available in stock sizes and shapes and can also be custom cut to specification.

Architectural cast stone is a very expensive stone product — crushed stone and cement cast into a molded shape.

Many natural stones are also available as veneers that can be used as facings on walls or floors.

■

DIVISION 5

METALS

05300 METAL DECKING

Metal decking is a kind of built-in formwork that is used as a base for a cast-in-place concrete floor. It supports the concrete while it's curing, and it is generally left in place. It is not typically used in residential construction, but can be.

05400 METAL FRAMING

Steel framing systems, long used in commercial construction, are now also being used in the building of healthful houses. This steel building framing is used for constructing the exterior wall envelope and interior partitioning. Loadbearing and nonstructural loadbearing fire-resistant steel studs, metal tracks, steel joists, and steel trusses can be faced with the same finishes as wood.

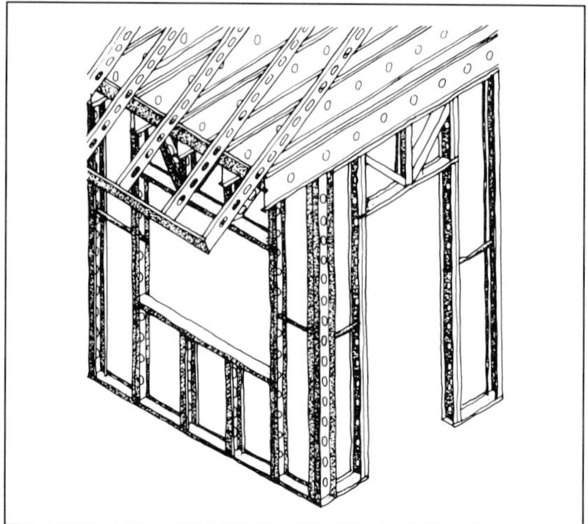

FIGURE 11: Light steel structural framing.

05500 METAL FABRICATIONS

Some items that are commonly built into a house are available as metal prefabricated units, such as stairways. Standard prefabricated metal stair units come with concrete steps, or you can also get metal spiral stairs with solid oak treads.

Handrails and railings are made from all types of metals: brass, aluminum, bronze, stainless steel, or enameled metal. You can buy them prefabricated or custom-built.

■

DIVISION 6

WOOD & PLASTIC

06100 ROUGH CARPENTRY

The wall studs and joists should be done entirely with solid woods such as kiln-dried pine or fir (these standard woods are generally not treated with preservatives). If any volatile substances are used in drying, they would dissipate during the typical six-week duration of the framing process.

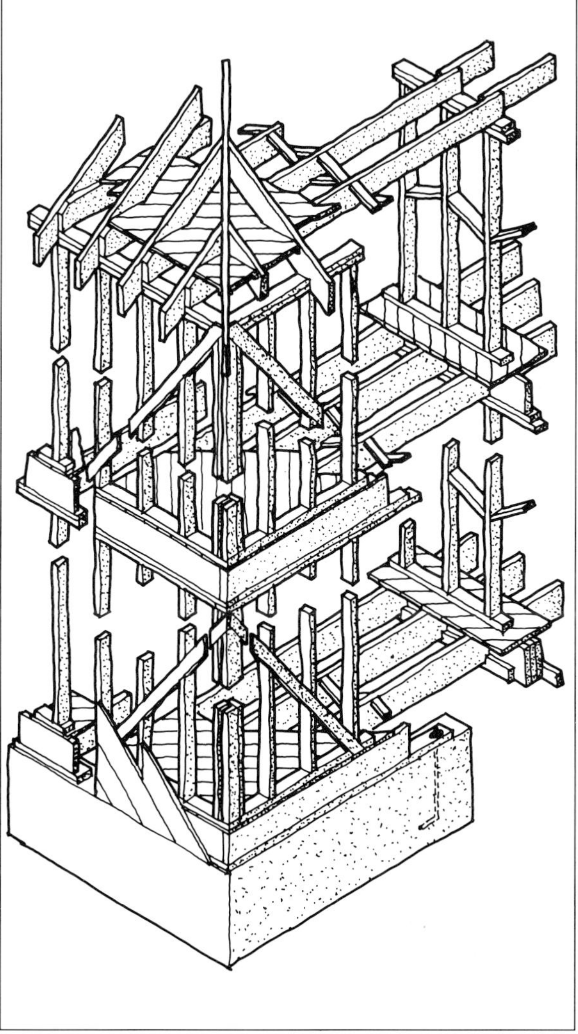

FIGURE 12: Wood frame construction.

For subflooring, exterior wall sheathing, and roof sheathing, use 1 x 6 solid pine or fir sheathing (in lieu of plywood) laid diagonally for strength (see Figure 12). Be sure to allow at least 1/2" space between sheathing boards for expansion.

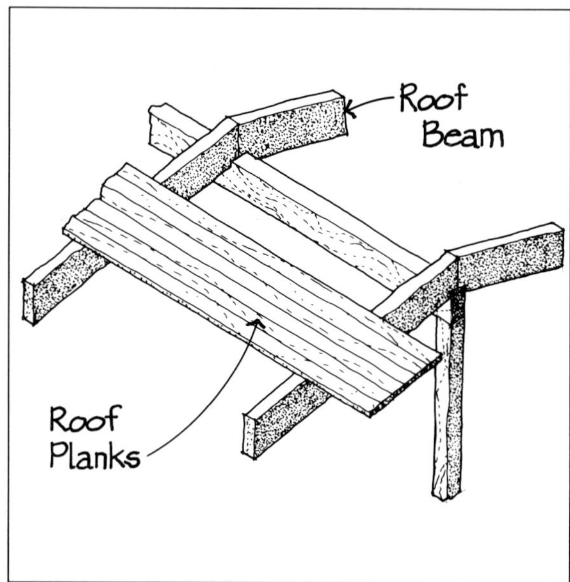

FIGURE 13: Heavy timber framing.

All blocking and supports to join members and anchor the framework should be of solid wood also.

Heavy timber construction, also known as post-and-beam construction, is another possible option. This involves framing with large pieces of timber and the use of wood decking (long, thick spans of tongue-and-grooved wood) for exposed flooring and roofing.

06150 WOOD-METAL SYSTEMS

Manufactured floor trusses with metal web components and 2 x 4 wood top and bottom cords are available, which speed the construction process.

06200 FINISH CARPENTRY

All exposed interior millwork such as paneling, window and door casings, base trim, moldings and other miscellaneous millwork should be done with unprocessed solid woods only.

PLASTIC LAMINATE

Plastic laminate is an inexpensive and relatively safe material that can be used for countertops. The health hazard comes not so much from the plastic laminate itself, but from the formaldehyde emitting from the sub-strate and the adhesive used in prefabricating countertops. This can be sealed by covering the bottom with a foil-backed building paper and foil tape, or with a sealant coating (see Section 09900).

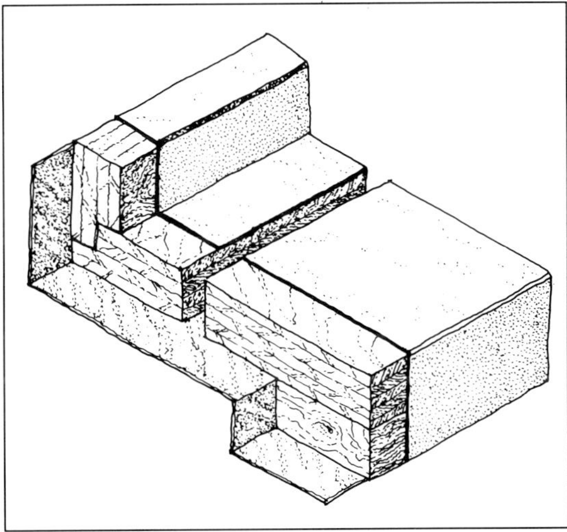

FIGURE 14: Plastic laminate countertop.

06300 WOOD TREATMENT

Treated woods are often used on outdoor decks and in construction where wood comes in contact with masonry or concrete. As the manufacturers' warnings for the hazards during installation of standard treated wood are dire, it is best to avoid pressure-, pest-, or preservative-treated woods and choose a natural wood treatment product that can be applied to untreated wood.

Brand name products:

Auro No. 111 Borax Wood Impregnation (see note on page 28). For preventive protection of wood from fungus and insects. Does not emit vapors.

Auro No. 115 Wood-Pitch Impregnation (see note on page 28). Resin-oil and beechwood distillate solution for outdoor wood protection and close-to-soil areas.

Livos Donnos Wood Pitch Impregnation (see note on page 28). A water-resistant exterior wood preservative. Protects against fungus and mildew.

06400 ARCHITECTURAL WOODWORK

All built-in cabinetry and wall paneling should be made of unprocessed solid wood only.

Brand name products:

Balzac Inc., 2750 Fenton Road, Gloucester ON Canada K1G 3N3 (U.S. distribution out of Flint Hill VA), 800/267-9792, 613/822-6557. High-quality oak wall paneling and wainscotting, fireplace mantles and pillars, prefabricated staircases.

Fieldstone, PO Box 109, Northwood IA 50459, 515/324-2114. Best-quality wood cabinets available.

Northeastern Hardwoods, PO Box 365, Salamanca NY 14779, 716/945-1510. Solid wood paneling and other hardwood and moldings.

Visador, PO Box 2090, Jasper TX 75951, 409/384-2564. Spiral wood staircases.

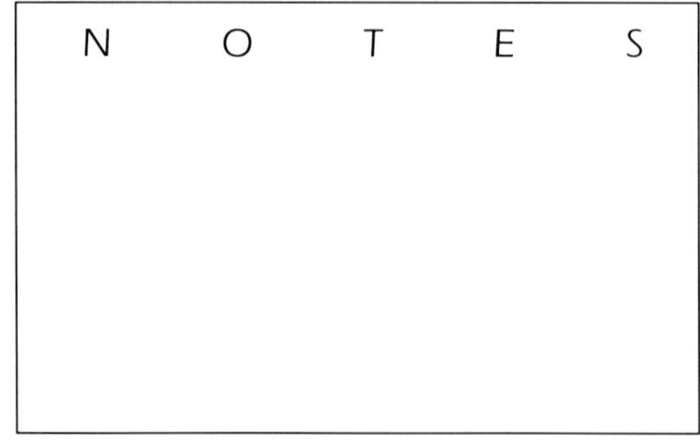

07150 DAMPPROOFING AND VAPOR BARRIER

DAMPPROOFING

Dampproof foundation walls and any wall that is adjacent to soil by using a treatment of concrete or mortar, or by applying a suitable sealer to retard the passage or absorption of water.

Brand name products:

Thoroseal Foundation Coating (Thoro Systems Products, 7800 NW 38th Street, Miami FL 33166). A cement-base heavy-duty dampproofing coating for exterior surfaces of concrete and masonry foundations below grade.

AFM Cem Bond (see note on page 28). A waterproof masonry sealer paint.

AFM Dyno Seal (see note on page 28). A waterproof and weather-resistant coating and sealer. AFM Dyno Flex mastic compound sealer, a thick finish paint, can be applied over Dyno Seal if desired.

Okon Waterbased Waterproofing (Okon Manufacturing Co., 6000 W 13th Avenue, Lakewood CO 80214, 303/232-3571).

Xypex Corporation Nontoxic Concrete Waterproofing by Crystallization™ (Xypex Corporation, Inc., 13731 Mayfield Place, Richmond BC Canada V6V 2G9, 604/273-5265).

VAPOR BARRIER

Vapor barrier paper can be used not only to block air and moisture, but also to block fumes from materials used in wall construction.

Brand name products:

Denny Foil Vapor Barrier (Denny Sales Corporation, Pompano Beach FL, 800/327-6616). Aluminum foil bonded to brown kraft paper (get the unperforated type).

DIVISION 7

THERMAL & MOISTURE PROTECTION

07200 INSULATION

Any type of rigid insulation used under slabs in the basement as perimeter insulation is acceptable where there is no interior exposure. For cavity insulation in walls and building insulation in the attic and crawlspaces, choose one of the healthful insulations listed here.

As late as 1948, architecture books were recommending crumpled aluminum foil for wall insulation (R-value 16), a good nontoxic product. Avoid urea-formaldehyde foam insulation, fiberglass batts, and polyisocyanate insulation boards.

Brand name products:

Air-Crete Foamed-In-Place Insulation (Air-Crete Inc., PO Box 380, Weedsport NY 13166, 315/834-6609). Cementitious magnesium-oxide foam, which hardens after application to inert (hard sponge). R-value slightly better than that of fiberglass batt insulation, according to literature.

Aluminum Insulation (Key Solutions, 7529 E Woodshire Cove, PO Box 5090, Scottsdale AZ 85258, 602/948-5150). Multi-layered aluminum foil insulation combining the principle of reflectivity with dead air spaces to provide an R-value comparable to standard insulations.

Perlite (Perlite Institute Inc., 600 S. Federal Street, Chicago IL 60605, 312/922-2062). Made from perlite, a naturally occuring mineral. Can also be used as loose fill insulation in cavities of concrete block.

07300 ROOFING

There are many healthful materials that can be used for roofing. Copper, terra cotta, clay or concrete roofing tiles, and slate are relatively expensive choices.

Wood shingles (cut to uniform size) and shakes (split to uneven sizes) are a natural choice. Specify that they be untreated to avoid fire retardant.

Terne metal and factory-applied baked-on enameled steel roofing are excellent options that are gaining wider acceptance. There are different types, colors, and profiles. Terne metal roofing is typically galvanized and unpainted and comes in a weathering type that weathers to a beautiful grey.

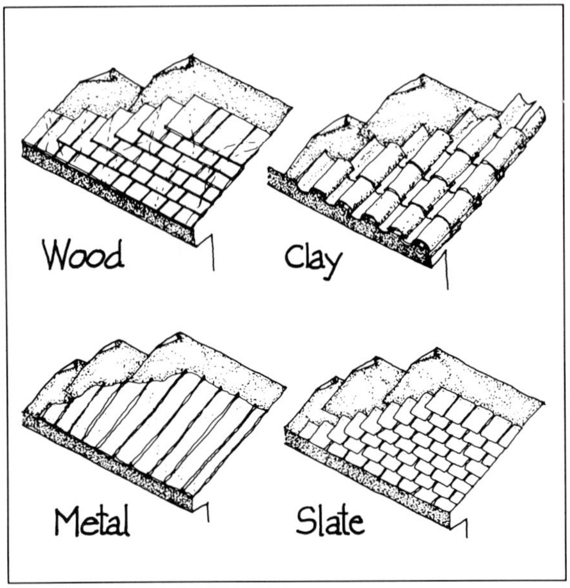

Wood
Clay
Metal
Slate

FIGURE 15: Various roofing tiles.

Asphalt and fiberglass shingles are acceptable but not preferred.

07400 PREFORMED ROOFING AND SIDING

Wood siding — cedar, redwood, cypress — has been used for hundreds of years to finish exterior walls. The wood has no preservatives, and naturally ages into a lovely driftwood grey. It could be sealed with a nontoxic sealant to retain its natural luster, stained with a natural stain, or painted with a natural paint.

Cedar shingle siding combines the look of shingles with the convenience of siding.

Porcelain-enameled siding, a preformed metal siding with a vitreous inorganic metal oxide coating bonded to metal by fusion at high temperature, and aluminum siding with baked-on finish could be used very effectively on houses of high tech or art deco styles.

Avoid plywood siding.

07600 FLASHING AND SHEET METAL

TERMITE SHIELDS

Termite shields come in many different sizes, for many uses. Some material supply houses carry rolls of sheet metal that can be used for termite shields. After unrolling the sheet metal over foundation walls, simply bend edges down at a 45° angle.

07800 ROOF ACCESSORIES

If you decide to install skylights, use metal frame skylights with glass glazing and avoid skylights with plastic glazing.

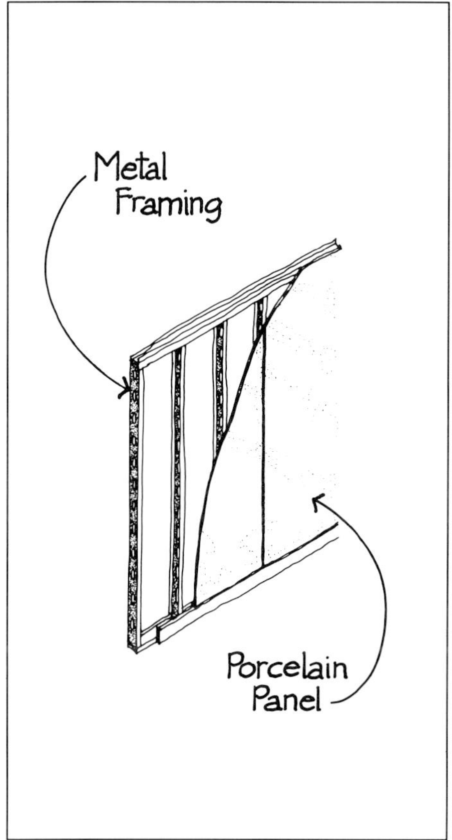

FIGURE 16: Porcelain panel wall construction.

07900 JOINT SEALANTS

On the exterior of the house you can use any silicone caulk. Look for brands that say "silicone" on the label.

For interior use, choose a caulk that emits a minimum of fumes.

Brand name products:

AFM Joint Compound (see note on page 28).

AFM Dyno Flex Caulking Compound (see note on page 28). A water-based, water-resistant rubber emulsion-type compound.

AFM Spackling Compound (see note on page 28).

Auro No. 386 Natural Joint Filler (see note on page 28).

Dow-Corning 100 Percent Silicone Rubber General Purpose Sealant (Dow-Corning Corp., Midland MI 48640).

Livos Vedo Spackling Compound (see note on page 28). An elastic caulking compound to repair and fill plastered surfaces, masonry and concrete. For interior use only.

Livos Anavo Oil-based Spackle (see note on page 28). Oil-based wood filler.

Livos Linseed Putty (see note on page 28). A water- and weather-resistant substance for sealing glass windows and cracks, gaps, and leaks around windows.

One Part Rubber Construction Sealant (Bostick, 2930 Turnpike Drive, Hatboro PA 19044, 215/674-5600).

N O T E S

DIVISION 8

DOORS & WINDOWS

08100 METAL DOORS AND FRAMES

Metal doors and frames are a very healthful product, but not aesthetically appealing. Perhaps you might limit their use to basement and garage doors, and other utilitarian areas.

Brand name products:

Benchmark (General Products Company, PO Box 7387, Fredricksburg VA 22404, 703/898-5700). An attractive metal bi-fold adjustable door.

08200 WOOD DOORS

Solid flush wood doors (with straight faces, no panels) are made with particleboard interiors and should be avoided. Those made with wood block cores are not much better because a lot of glue is used to put the blocks together. Doors or frames made from "finger-joints" (small pieces of wood glued together in a pattern that looks like interlaced fingers) also use a lot of glue in construction.

Choose wood panel doors or french doors (double wood doors with glass panels) which are made from solid, unprocessed wood. If you are on a budget,

FIGURE 17: Typical door styles.

solid wood doors with interesting designs may be purchased inexpensively at your local building salvage yard.

08500 METAL WINDOWS

Metal frame windows are a healthful and aesthetic product for the new home. Generally, whatever weatherstripping and sealant materials that are used in the construction of a metal window are much less of a problem than the preservative that may be used in the construction of a wood window.

08600 WOOD WINDOWS

Wood windows may appear to be natural and healthful; however most manufacturers use preservatives and moisture-repellents on interior parts of windows.

Wood windows can be used successfully if a sealant is applied that will block any offending fumes from the preservatives (see Section 09900).

If you are on a budget, beautiful old wood windows with no preservatives can frequently be purchased at salvage yards and be stripped and refinished.

08700 WEATHERSTRIPPING

Use a felt strip bound with metal or a hard vinyl strip. Integrate weatherstripping at the base of the door with the door threshold.

N O T E S

■

DIVISION 9

FINISHES

09100 METAL SUPPORT SYSTEMS

Metal non-loadbearing framing systems, ceiling suspension systems, and furring systems are used as a hanging system for suspending drywall on ceiling and walls.

09200 LATH AND PLASTER

Plaster over metal mesh or gypsum board lath gives a beautiful textured wall finish. In the old days wood lath was used, but to use wood nowadays is an unnecessary expense.

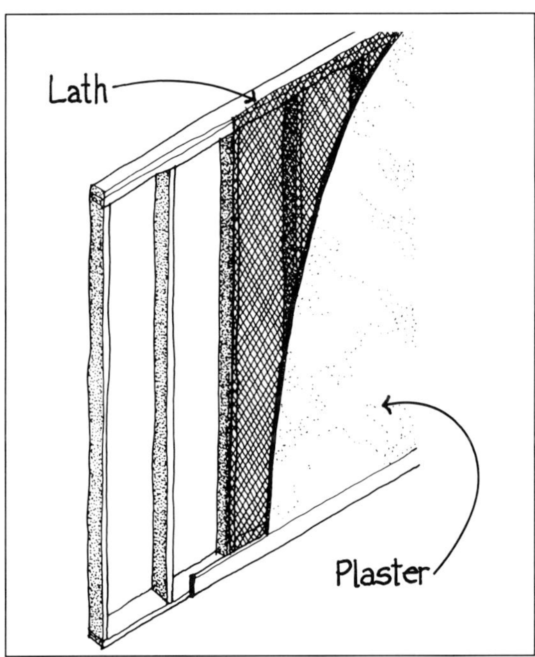

FIGURE 18: Plaster wall.

09250 GYPSUM WALLBOARD

Any gypsum wallboard is acceptable. If you are using a foamed-in-place insulation such as Air-Crete, you must use aluminum foil-backed gypsum wallboard, or the insulation absorbs into the drywall and you've got a big mess.

An asbestos-free joint compound should be used when applying joint tape. Most joint compounds contain formaldehyde and other volatile substances so it is important to allow the joint compound to dry thoroughly before it is painted so the chemicals can volatilize completely.

Avoid using kerosene or gas heaters to dry joint compound because the combustion byproducts from the heater can be absorbed by the wallboard and joint compound. This is very important. Use turbo high-velocity air dryers instead, which can be rented at local machine and equipment rental stores.

Brand name products:

AFM Joint Compound (see note on page 28).

Murco M-100 HiPo Asbestos-free Drying Tape (Murco Wall Products, Fort Worth TX, 817/626-1987).

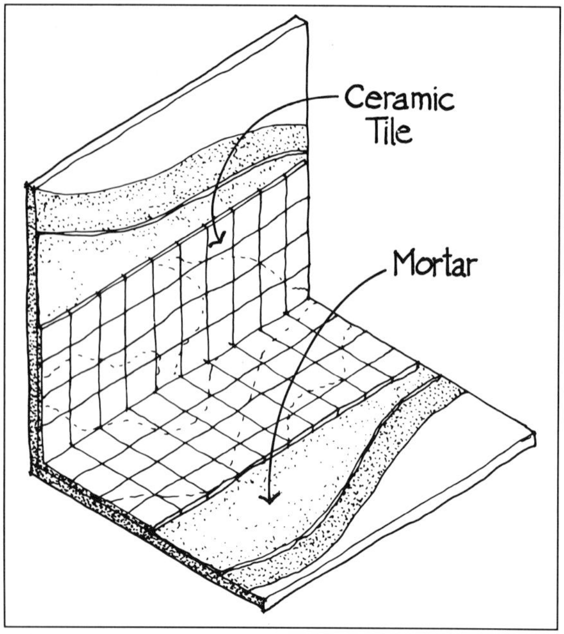

Ceramic
Tile
Mortar

FIGURE 19: Ceramic tile.

09300 CERAMIC TILE

To avoid the common practice of installing ceramic tile using noxious adhesives, the tile may be installed in mortar, as follows: First put down cleavage paper (non-asphalt-impregnated) over the concrete or wood sheathing, next metal lath reinforcing, covered with a scratch coat of mortar (you don't need the recommended latex additives) and a $1\frac{1}{4}$" thick setting bed of mortar. Lay your tile, and fill in with a grout that does not contain fungicides or latex additives. The same installation can be used for walls and countertops.

Brand name products:

GROUT

AFM Tile Grout (see note on page 28).

GROUT SEALANT

AFM Water Seal (see note on page 28). A nontoxic sealant that will also block toxic fumes emitting from grout additives.

UNDERLAYMENT

Substrate '500' (Eternit Inc., Village Center Drive, Reading PA 19607, 215/777-0100). A strong, lightweight, noncombustible, highly water-resistant base for ceramic tile made from concrete and cement, reinforced with treated fiberglass mesh. Does not contain asbestos. Not affected by moisture, so it's a good material to use under countertops and around tubs.

09400 TERRAZZO

Terrazzo is an expensive but healthful flooring material, either cast-in-place marble or pre-cast and ground smooth.

09550 WOOD FLOORING AND FINISHES

No. 2 Common Red Oak with all of its wormholes and knots provides a very beautiful finish, in my opinion. For a more refined look, choose better grades of oak.

Prefabricated floors made of different types of woods (oak, walnut, and assorted hardwoods) are available with interesting patterns. Tongue-and-grooved wood flooring is generally nailed diagonally so the number of exposed nail holes is limited. Do not fill holes with wood filler, just countersink nails so they appear to be wormholes. Choose a hardwood floor that can be nailed as opposed to one that needs to be applied with adhesive (if adhesive type must be used, plain white glue is very effective).

Your local lumberyard or millwork and woodworking shop should have a good selection of hardwood flooring products. Be sure to avoid hardwood floor tiles that are glued together or have foam backings.

Brand name products:

WOOD STRIP FLOORING (HARDWOOD FLOORING)

Bruce Hardwood Floors, 16803 Dallas Parkway, Dallas TX 75248, 214/931-3100.

Bangkok Industries Inc., 4562 Worth Street, Philadelphia PA 19124, 215/537-5800.

Harris-Tarkett Prefinished Solid Hardwood Floors, 383 East Maple Street, Johnson City TN 37601-0300, 615/928-3122.

Hoboken Wood Floors, East Rutherford NJ, 201/933-9700. Beautiful inlaid flooring.

WOOD FLOOR FINISHES

AFM Hard Seal (see note on page 28). A medium-gloss, nontoxic finish.

AFM Polyuraseal (see note on page 28). A mar-resistant, water-resistant nontoxic finish especially for floors.

FIGURE 20: Wood flooring patterns.

Hydroline Wood Floor Finish (Basic Coatings, 2137 Sunset Road, PO Box 677, Des Moines IA 50303, 800/247-5471, in Iowa 800/622-8240). Nonflammable, clear, "almost odorless" coating dries in two hours. Glowing finish with no waxing or stripping.

Livos Meldos Hard Sealer (see note on page 28). A water-resistant finish with strong surface hardening capability.

Livos Natural Resin Floor and Furniture Lacquer (see note on page 28). Strong finish with medium to high gloss for interior use on wood, stone, and metal. Can be tinted.

WOOD FLOOR WAXES

AFM All-Purpose Polish and Wax (liquid) (see note on page 28).

Auro No. 171 Floor Wax (see note on page 28). Water-repellent, made from beeswax and plant waxes.

Auro No. 431 Beeswax Floor Care (see note on page 28). Organic emulsion for care of waxed floors, forms water-repellent protective coating.

Livos Bilo Floor Wax (see note on page 28). Can be buffed to high satin gloss. For use on wood, stone, brick, terra cotta, and linoleum.

WOOD FLOOR ADHESIVES

Auro No. 384 Natural Parquet Adhesive (see note on page 28). Natural resin glue.

09600 STONE AND BRICK FLOORING

Slate and brick pavers make good healthful floors.

09650 RESILIENT FLOORING

Avoid vinyl linoleum floors because they emit fumes of carcinogenic vinyl chloride. If linoleum is desired, use natural linoleum.

Brand name products:

LINOLEUM

Forbo North America, PO Box 32155, Richmond VA 23294, 800/233-0475, 804/747-3714. "Plain Battleship Linoleum" made from linseed oil, pine resins, wood flour and jute. Solid colors and marbleized patterns.

ADHESIVES

Auro No. 383 Natural Linoleum Glue (see note on page 28). Natural resin glue.

Livos Melino Carpet and Linoleum Adhesive (see note on page 28).

09680 CARPETING

Choose a natural fiber carpet, or another type of flooring. If you must have a synthetic carpet for economic reasons, you can minimize emissions from the carpet by applying AFM Carpet Guard (see note on page 28), an odorless penetrating sealant that makes carpets water resistant as well.

Brand name products:

CARPETS

Carousel Carpet Mills, 1 Carousel Lane, Ukiah CA 95482, 707/485-0333. Carries a wide selection of natural fiber carpets and can custom make almost anything you want. Most of their carpets are wool, but they also have a unique handloomed natural color linen and cotton woven carpet without latex backing. Expensive.

Dellinger, PO Drawer 273, Rome GA 30161. A less expensive cotton pile carpet. Can be dyed to match any custom color.

Selfhelp Crafts, 704 Main Street, Box L Dept HH, Akron PA 17501, 717/859-4971. Imported pesticide-free oriental, dhurrie, and Persian rugs.

CARPET ADHESIVE

Auro No. 385 Natural Carpet Glue (see note on page 28). Natural resin glue.

Livos Melino Carpet and Linoleum Adhesive (see note on page 28).

UNDERLAYMENT

Homasote 440 CarpetBoard (Homasote Co., PO Box 7240, West Trenton NJ 08628-0240). Has insulation, sound deadening, and cushioning qualities. Can be installed over concrete or wood. "No asbestos or urea formaldehyde additives".

09900 PAINTS AND FINISHES

If you are on a budget and do not want to invest in natural or nontoxic paints for your entire house, you can get around some of the odor problems with conventional paint by using a water-based latex paint and then "baking" the interior of the house before moving in. Just put a heat source in the house (such as an electric space heater or two), close all the doors and windows, and let it heat up. After about a week, the paint will be cured about the same as if it had been drying for six months under normal conditions.

Brand name products:

PAINT

AFM Safecoat Paint (see note on page 28). A flat water-based copolymer emulsion wall paint for interior use. Comes in two shades of white, but can be tinted with universal colors that can be purchased at any paint store.

AFM Safecoat Enamel Paint (see note on page 28). A semi-gloss enamel for interior use. Comes in two shades of white.

AFM All-Purpose Enamel (see note on page 28). A high-gloss interior/exterior paint. Comes in two shades of white.

AFM Cem Bond (see note on page 28). A masonry sealer paint that also functions as an excellent exterior paint. Comes in two colors.

Auro No. 321 Natural Resin Wall Paint, White (see note on page 28). Indoor semi-gloss paint.

Auro No. 325 Natural Chalk Wall Paint (see note on page 28). Inexpensive ready-mix indoor wall paint.

Livos Vindo Enamel Paint (see note on page 28). Oil-based enamel paint for interior and exterior use.

Livos Natural Resin Wall Paint (see note on page 28). Fast-drying wall paint for interior use. Can be tinted.

Livos Amellos Solvent Free Paint (see note on page 28). For interior and exterior wood and stone surfaces.

Livos Albion White Wash Paint (see note on page 28). Interior use whitewash.

Old-Fashioned Milk Paint (The Old-Fashioned Milk Paint Company, PO Box 222, Groton MA

04150, 617/448-6336). Milk-based paint. Good for walls but not suitable for high-moisture areas such as bathrooms because it molds easily.

PRIMER

AFM Safecoat Primer Undercoater (see note on page 28). For interior and exterior use.

Auro No. 121 Natural Resin-Oil Primer (see note on page 28). All-purpose primer for interior and exterior applications.

Livos Dubno Primer Oil (see note on page 28). Interior/exterior primer for wood, absorbent stone, slate and brick.

Livos Menos-Primer (see note on page 28). For wood, stone, and noncorrosive metals.

RADIATOR PAINT

Auro No. 237 Natural Resin Radiator White (see note on page 28).

Livos Aidu Radiator Paint (see note on page 28).

STRIPPERS

AFM Stripper 66 (see note on page 28). A versatile, cold-cleaning and film-stripping water-based liquid solvent compound blend.

WOOD FINISHES

AFM Hard Seal (see note on page 28). A medium-gloss finish that can be used on woodwork.

AFM Shingle Protek (see note on page 28). A penetrating sealant for exterior wood shingles.

Auro No. 131 Natural Resin-Oil Varnish (see note on page 28). An all-purpose, transparent, tintable varnish for outdoor wood protection.

Auro No. 213 Clear Semi-Gloss Shellac (see note on page 28). Quick-drying semi-gloss lacquer for wooden indoor surfaces.

Auro No. 235 Natural Resin-Oil Lacquer, White (see note on page 28). Semi-gloss lacquer for interior and exterior use.

Auro No. 240 Natural Resin-Oil Lacquer, Colored (see note on page 28). Semi-gloss lacquer for interior and exterior application in eight colors.

Livos Kaldet Resin and Oil Finish (see note on page 28). A satin to semi-flat water-resistant finish for interior and exterior use on architectural woodwork. Can be tinted.

Livos Meldos Hard Sealer (see note on page 28). A water-resistant finish with strong surface-hardening capability for wood cabinets, doors, and windows.

WOOD STAINS

AFM Safecoat Waterbase Wood Stain (see note on page 28). For interior and exterior use. Comes in four colors.

Livos Bela Wood Stain (see note on page 28). A glaze-like finish for interior architectural woodwork. Cannot be used in high-moisture areas such as floors and bathrooms.

VAPOR BARRIER SEALANTS

These sealants will block formaldehyde fumes from particleboard or plywood as well as providing a durable finish.

AFM Water Seal (see note on page 28). Will seal in about 85% of formaldehyde fumes.

Livos Trebo All Purpose Shellac (see note on page 28). Seals up to 80% of formaldehyde fumes.

09950 WALL COVERING

Check your local wallpaper supply store for "paper"-type wallpaper without vinyl finish.

Brand name products:

AFM Wallpaper Adhesive (see note on page 28).

Auro No. 389 Natural Wallpaper Glue (see note on page 28).

Livos Lavo Wallpaper Paste (see note on page 28).

■

DIVISION 10

SPECIALTIES

Specialties are not generally used in residential construction.

■

DIVISION 11

EQUIPMENT

11450 APPLIANCES

The basic rule for specifying appliances is that they be electric — ranges, dryers, and water heaters. In many areas an all-electric house qualifies for a price break from your local utility company, and they have the capability to be powered by home-generated energy.

Brand name products:

REFRIGERATORS

Sub-Zero Refrigerators and Freezers (Sub-Zero Freezer Co., PO Box 4130, Madison WI 53711, 608/271-2233). Built-in refrigerators with glass interiors and exterior panels of virtually any material.

Whirlpool and General Electric have refrigerators that have porcelain interiors.

COOKTOPS

Sears Kenmore, General Electric and Kitchen-Aid have induction ranges. Since they use a glass cooking surface and only the pans are heated through conduction, foodstuffs do not burn onto the elements and emit odors.

METAL KITCHEN CABINETS

Davis Kitchen Products, 111 Beeson Street, Dowagiac MI 49047, 616/782-5188. Very elegant European-style kitchen cabinets.

■

DIVISION 12

FURNISHINGS

12500 WINDOW TREATMENT

Preferably, do as little to a window as possible. When designing your house, take privacy into consideration when placing windows, and allow in light and contact with nature.

If you wish to cover or decorate your windows, use draperies or curtains made from all cotton, wool, silk, or linen; or shutters, woven woods, wooden blinds, mini-blinds, grasscloth shades, natural-fiber fabric shades, or rice paper shades.

12600 FURNITURE

Choose furniture that is made from natural materials such as wood or metal and finish them with nontoxic or natural finishes (see Section 09900).

Custom-made upholstered furniture can be stuffed with cotton, feathers, wool, or kapok and covered with a variety of untreated natural upholstery materials made from cotton, silk, linen, wool or leather.

■

DIVISION 13

SPECIAL CONSTRUCTION

13150 SWIMMING POOLS AND HOT TUBS

If you are planning a swimming pool or hot tub (especially if it is to be indoors), you should be aware of the hazard of high concentrations of chloramine and chloroform compounds that form as byproducts of using chlorine products to purify the water.

There is now revolutionary new technology in the water sanitizing industry with which water can be sanitized without the hazardous side effects of chlorine. The best of these systems for swimming pools and hot tubs, to my knowledge, is described below.

Brand name products:

O3UV Systems for Swimming Pools and Spas (O3UV Systems, Inc., 8033 Sunset Boulevard, Suite 595, Los Angeles CA 90046, 213/653-6040). A unique water sanitizing system utilizing ozone and ultraviolet light. It produces clear, odorless, biologically safe water.

13980 WIND, HYDRO AND PHOTOVOLTAIC ENERGY SYSTEMS

When planning your house, you might want to consider incorporating a system for homemade electricity, by harnessing the energy of wind, water (micro-hydro) or sun (photovoltaics).

As the technology advances, wind power becomes more promising as a reliable source of energy for the homeowner. You must first ascertain that the wind resources available at your site are sufficient — wind speed should average 12 to 15 miles per hour and the site should be free of nearby wind obstacles to maximize energy potential. State and local regulations will need to be consulted when considering placement. Bear in mind that maintenance on a wind machine can be high.

For individuals who have a stream flowing through their property, micro-hydro systems can be an economical and efficient source of electricity, often providing all of the household electrical needs. A wide variety of streams will serve the purpose, from a fast-moving creek to a lazy, meandering river. The power production relies on two variables: flow (the volume of water available) and head (the vertical distance the water drops). The water passing through your property, however, may be subject to rules and regulations that either limit or prohibit its use, and various state and local agencies may have jurisdiction over a micro-hydro project.

The technology of converting sunlight into energy is clean, reliable, and nonpolluting. Systems are modular and can be added to as the need arises. Because photovoltaics have no moving parts, they are virtually maintenance free. They are currently quite expensive, although the cost is coming down and many predict that photovoltaics will soon be fully economical for the average homeowner. The main disadvantage to using solar power is that systems only produce DC current, which requires either converting to DC appliances or purchasing an inverter to convert to AC current.

The U.S. Department of Energy publishes an excellent booklet called *An Introduction to Small-Scale Wind, Hydro and Photovoltaic Systems* which can be ordered through the U.S. Government Printing Office, Washington DC 20402.

■

DIVISION 14

CONVEYING SYSTEMS

Conveying systems, such as dumbwaiters and residential elevators, are typical types of conveying systems used in residential construction.

■

DIVISION 15

MECHANICAL

15050 BASIC MATERIALS AND METHODS (FOR PLUMBING)

WATER SUPPLY SYSTEM

In no instance should there be any leaded solder used in any part of the water supply system. Copper is the recommended piping material. Avoid polyvinyl chloride (PVC) and polybutylene piping.

WASTE DRAIN SYSTEM

Use cast iron piping with "no-hub" connectors. Many plumbers with commercial building experience use this type regularly.

15450 PLUMBING FIXTURES AND TRIM

Visit a local plumbing supply center that carries a good selection of reputable brand plumbing fixtures, and choose vitreous china sinks and toilets, porcelain-enameled cast iron bathtubs (painted steel chips off), and fixtures made of stainless steel, porcelain, enameled steel, or enameled cast iron.

15550 WATER FILTRATION

Today, houses in most areas need to provide some sort of water purification system. If you have tap water you probably need to filter out volatile trihalomethanes and fluoride; if water comes from a well or spring it may need treatment to make it bacteriologically safe.

It is good to consider what type of water treatment will be needed in your house before you build, so that you can easily install appropriate equipment.

To find out what's in your water, call city hall, your local water district office, or your local department of health services and ask them the following questions:

❒ Where does your water come from? a reservoir? groundwater?

❒ What type of pipe is used to transport the water? Does it add lead, asbestos, or vinyl chloride?

❒ Is chlorine used to disinfect the water? or chloramine? or some other chemical?

❒ Is your water fluoridated?

Other questions to consider are:

❒ Is there agriculture in your area that would result in excessive pesticide or nitrogen fertilizer runoff? Or are there factories nearby producing industrial waste?
❒ Where is your city dump in relation to municipal water supplies? Could hazardous materials from the dump be leaching into your water?

For the purposes of removal, water pollutants fall into four groups:

❒ **Particulates** (minute bits of material that do not dissolve in water) — asbestos, arsenic, heavy metals (aluminum, cobalt, chromium, nickel, mercury, lead, cadmium, manganese, silver), rust, dirt, sediment, etc.
❒ **Dissolved solids** (solid materials that decompose in water) — fluoride, nitrates, sulfates, salts, etc.
❒ **Volatile chemicals** (nonparticulate substances that evaporate) — chlorine, chloramine, chloroform, chlorinated hydrocarbons, pesticides (DDT, dialdrin, lindane, heptachlor, etc.), PCBs, benzene, carbon tetrachloride, trichloroethylene, xylene, toluene, etc.
❒ **Microorganisms** (microscopic plant and animal life) — bacteria, viruses, etc.

To remove these pollutants, there are basically three types of units:

❒ **Activated carbon** — removes volatile chemicals.
❒ **Reverse osmosis** — removes particulates and dissolved solids.
❒ **Distillation** — removes particulates, dissolved solids, and microorganisms.

Chances are, you will need a combination of these types. The preferred arrangement for typical municipal water would be to have a whole house activated carbon filter installed at the "point of entry" to remove volatile chemicals and a distiller installed to make drinking water free from particulates, dissolved solids, and microorganisms. Of course, there are many other systems available that have varying degrees of effectiveness and may be appropriate for individual needs.

Brand name products:

Rain Crystal Water Distillers (Scientific Glass, 113 Phoenix NW, Albuquerque NM 87107, 505/345-7321). A Pyrex glass distiller for home use, with optional activated carbon.

Seagull IV (General Ecology, 151 Sheree Boulevard, Lionville PA 19353, 215/363-7900). Stainless steel units filled with enhanced adsorptive material, similar to activated carbon.

Water Safe WS/RO5 (Coast Filtration, 142 Viking Avenue, Brea CA 92621, 714/990-4602). An effective reverse osmosis/activated carbon unit. Their Water-Safe TOC-200 is an economical activated carbon block unit.

15600 HEATING, VENTILATION AND AIR CONDITIONING

HEATING AND COOLING SYSTEMS

There are many systems for providing heating and cooling for our homes. The basic systems are:

❒ **Circulating hot water** (hot water can be heated with electricity or a remote gas boiler).

❒ **Electric radiant heat** (oil- or water-filled radiators and baseboard units).

❒ **Central air conditioning and heating** (the air can be heated and cooled using gas, oil or heat pump. Again, the gas and oil furnace can be located in a remote, well-ventilated room.)

More and more, people are considering systems which rely on the natural heating and cooling abilities of earth, water and sun, as an alternative to using nonrenewable fossil fuels:

❒ **Earth and water coupled systems** use the thermal energy of the earth and the earth's ground and surface waters to heat and cool your home:

1) Ground water-coupled system with heat pump.
2) Surface water-coupled system with heat pump.
3) Water-coupled system with heat exchanger.
4) Earth coil with heat pump.
5) Earth tubes.

Each of these systems have distinct advantages and disadvantages, all of which need to be determined for appropriateness and cost-effectiveness.

❐ **Solar heating systems:** Active or passive.

HEAT RECOVERY VENTILATORS

Heat recovery ventilators (also known as air-to-air heat exchangers) are ideal for creating a comfortable, healthy indoor environment. Heat recovery ventilators (HRVs) pull stale, warm air from the house and transfer the heat in that air to the fresh air being pulled in the house. HRVs do not produce heat themselves, but instead preserve heat generated from other sources, while at the same time bringing fresh air in from outside.

Residential HRVs come in two basic types:

❐ **Through-the-wall units** are about the size of a room air conditioner and are usually used for one or two rooms.

❐ **Central units** are ducted and used for the whole house.

While we have tightened our homes to conserve energy, we have reduced air infiltration to unhealthy levels — our buildings no longer "breathe". An HRV can do an excellent job of supplying ventilation while at the same time conserving energy; however, the ventilation provided by an HRV should not be expected to sufficiently dilute high levels of indoor air pollutants by itself.

Brand name products:

Airxchange Inc., 401 VFW Drive, Rockland MA 02370, 617/871-4816.

Des Champs Laboratories Inc., Box 440, 17 Farinella Drive, East Hanover NJ 07936, 201/884-1460.

EER Products, Inc., 3536 E. 28th Street, Minneapolis MN 55406, 612/721-4231.
Mountain Energy and Resources Inc., 15800 W Sixth Avenue, Golden CO 80401, 303/279-4971.

AIR FILTRATION

Many airborne pollutants can be reduced by using an air filtration device integrated with the central heating system.

The most effective type of filter for removing particulate pollutants (such as pollen, dust, animal hairs, and molds) is the HEPA (High Efficiency Particulate Accumulator) filter. Its unique design causes it to actually become even more efficient with use.

To remove volatile gas pollutants (such as formaldehyde, chloroform, acetone, etc.), activated carbon must be used.

Filters can be incorporated into a central system to purify the air throughout the house, or small portable filters can be placed in each room to reduce pollutant levels.

It is important to remember that under typical conditions air filters can significantly *reduce* levels of pollutants, but do not generally provide 100 percent removal without sealing off an area and recycling the same air through the filter repeatedly.

PLEASE NOTE: Air filters are *not* a substitute for reducing pollutant levels at the source. I am *not* recommending that you buy an air filter as a means of making toxic materials acceptable. It is most appropriate to use air filters when all major sources of indoor air pollution have been removed and significant levels of pollutants are still present because of inferior outdoor air quality, or if exposure to a particularly harmful pollutant cannot otherwise be avoided. *The preferred method is to build with healthful materials in a clean-air area.*

Brand name products:

Aireox Clean Aire series (Aireox Research Corporation, 11015 Whitford Avenue, Riverside CA 92505, 714/689-2781).

Allermed VH300 or Chairman 800 (Allermed Corporation, 631 J Place, Plano TX 75074, 214/248-0782).

Foust 160 series or 400 series (E.L. Foust Company, PO Box 105, Elmhurst IL 60126, 312/834-4952).

King-Aire Freedom, Champion 300, or Guardian (King-Aire, PO Box 149, Carmel IN 46032, 317/845-1170

N O T E S

DIVISION 16

ELECTRICAL

16050 BASIC MATERIALS AND METHODS

WIRING

The initial installation of plastic coated electrical wiring outgasses fumes that are very noxious. This presents more of a problem for the electrician than the home occupants; however, if you are concerned about the residual outgassing of this material, you may consider specifying metallic sheathed wiring or electrical metal flexible tubing.

RECEPTACLES (OUTLETS) AND COVER PLATES

Electrical contractors have been using plastic receptacle housings and cover plates for many years now. It may seem ludicrous to some to revert to metal receptacle housings and steel, brass, enameled, or ceramic cover plates, but if your aim is to reduce the total chemical load in your house, this additional step is very inexpensive.

16500 LIGHTING

The electrical specifications usually require the contractor to include in his bid a certain specified allowance for lighting fixtures, giving the homeowner an opportunity for selection.

When choosing the lighting fixtures, look for fixtures that have little or no plastic, especially in the sockets which will be heated by the hot bulb. There are many fixtures made from metal, glass, or solid wood.

FULL SPECTRUM LIGHTING

Mother Nature provides a fully balanced spectrum of colors in sunlight. Most artificial lighting, however, gives only a narrow spectrum with emphasis on blues and yellows.

To find out about long-life full-color spectrum light bulbs, contact your local electrical lighting supply center or dealer.

■

SUGGESTED READING

Barbara, E.J., Montanaro, A., *Tight Building Syndrome,* Immunology and Allergy Practice. VIII/3. 17—31, March 1986.

California State Department of Consumer Affairs. *Clean Your Room! A Compendium Describing a Wide Variety of Indoor Pollutants and Their Health Effects, and Containing Sage Advice to Both Householders and Statespersons in the Matter of Cleaning Up.* Sacramento: California State Department of Consumer Affairs, February 1982.

Dadd, Debra Lynn. "Building A Healthy Home," *Everything Natural,* September/October 1986.

Dadd, Debra Lynn. *Nontoxic & Natural: How to Avoid Dangerous Everyday Products and Buy or Make Safe Ones.* Los Angeles: J.P. Tarcher, 1984.

Dadd, Debra Lynn. *The Nontoxic Home: Protecting Yourself and Your Family from Everyday Toxics and Health Hazards.* Los Angeles: J.P. Tarcher, 1986.

Gammage, R.B., Kaye S.V., *Indoor Air and Human Health.* Lewis Publishers, Chelsea, MI. 1985.

Godish, Thad. *Air Quality.* Lewis Publishers, Chelsea, MI. 1985.

Greenfield, Ellen J. *House Dangerous.* Vintage Books, New York. 1987.

Meyer, Beat. *Indoor Air Quality.* Addison-Wesley, Reading, Mass. 1983.

National Research Council/National Academy of Science. *Indoor Pollutants.* Washington DC: National Academy Press, 1981.

Pearson, David. *The Natural House Book.* Simon & Shuster Inc. London, England, 1989.

Pfeiffer M.D., Guy O., Nikel, Casmir M. *The Household Environment and Chronic Illness.* Charles C. Thomas, Springfield, IL. 1980.

Rousseau, David. *Your Home, Your Health, and Well-Being.* Hartley & Marks, Vancouver, B.C. Canada, 1988.

Ulsamer, Andrew G., and Beall, James R. "Overview of Health Effects of Formaldehyde," *Hazard Assessment of Chemicals: Current Development.* United States Consumer Product Safety Commission, 1984.

Ulsamer, Andrew G., and Beall, James R. "Toxicity of Volatile Organic Compounds Present Indoors," *The Bulletin of the New York Academy of Medicine.* Vol. 57, No. 10, December 1981.

United States Environmental Protection Agency. *Radon Reduction Methods: A Homeowner's Guide.* United States Government Printing Office, August 1986.

Verolia, Carol. *Healing Environments.* Celestial Arts Publishers, Berkeley, CA., 1989.

Zamm, Alfred, Gannon, R. *Why Your House May Endanger Your Health.* Simon and Schuster, New York. 1980.